A Retractable Guidance System for Mine Shaft Hoists

A Retractable Guidance System for Mine Shaft Hoists

Paweł Kamiński
AGH University of Science and Technology, Kraków, Poland

CRC Press is an imprint of the
Taylor & Francis Group, an **informa** business

First published 2022
by CRC Press/Balkema
Schipholweg 107C, 2316 XC Leiden, The Netherlands
e-mail: enquiries@taylorandfrancis.com
www.routledge.com – www.taylorandfrancis.com

CRC Press/Balkema is an imprint of the Taylor & Francis Group, an informa business

Library of Congress Cataloging-in-Publication Data
A catalog record has been requested for this book

ISBN: 978-1-032-11275-6 (hbk)
ISBN: 978-1-032-12897-9 (pbk)
ISBN: 978-1-003-22675-8 (ebk)

DOI: 10.1201/b22695

Typeset in Times New Roman
by codeMantra

Contents

Preface *vii*
About the author *ix*
Acknowledgments *xi*

1 State of the art **1**
1.1 Leon IV shaft *1*
1.2 Rope guidance of conveyances *15*
References *34*

2 Description of the retractable guidance system **37**
2.1 Characteristics of the retractable guidance system *37*
2.1.1 Bottom transom stabilizing 47
2.1.2 Top transom stabilizing 48
2.1.3 Retracting of the guides 48
2.2 Calculations *50*
2.2.1 Bottom section calculations 51
2.2.2 Top section calculations 52
References *62*

3 Measurements **65**
3.1 Measurements of peak forces acting on stiff guides from the cage *65*
3.1.1 Bottom transom frontal forces 70
3.1.2 Top transom frontal forces 70
3.1.3 Bottom transom side forces 74
3.1.4 Top transom side forces 77
3.1.5 Conclusions 79
3.2 Measurements of peak forces acting on the retractable guides *80*
3.2.1 Measurement schedule 80
3.2.2 "Bottom transom" stage 81
3.2.3 "Top transom" stage 85
3.2.4 Diagrams and markings of measured forces 89
3.2.5 Bottom transom frontal forces 89
3.2.6 Top transom frontal forces 91

3.2.7 Bottom transom side forces 93
3.2.8 Top transom side forces 96
3.2.9 Physical verification of measured forces 98
3.2.10 Stress in the bridle hangers of the man-material cage caused by the bottom section of the retractable guidance system 98
3.2.11 Stress in the bridle hangers of the man-material cage caused by the top section of the retractable guidance system 101
3.2.12 Physical verification of measured stress 102
3.2.13 Comparison of forces measured for stiff guides and for the retractable guidance system on the level 960 m in the Leon IV shaft 104
3.2.14 Comparison of forces measured for the retractable guidance system and existing legislation in terms of safety and Polish Standard PN-G-46227. Mine shafts: Chairing: Requirements 105
3.2.15 Evaluation of measured forces acting on the retractable guidance system 106
3.2.16 Evaluation of stress in the bridle hangers of the man-material cage caused by the retractable guides 107
3.2.17 Final conclusions of the retractable guidance system measurements 109

3.3 Mechanical analysis of construction of the retractable guidance system 110

3.3.1 Diagrams and calculations of forces acting on the bottom section of the retractable guidance system during its movement from resting to working position 111
3.3.2 Diagrams and calculations of forces acting on the top section of the retractable guidance system during its movement from resting to working position 112
3.3.3 Diagrams and calculations of forces acting on the bottom section of the retractable guidance system during its operation 114
3.3.4 Diagrams and calculations of forces acting on the bottom section of the retractable guidance system during its operation 116

References 118

4 Summary 119

Preface

Rope guidance of conveyances is in common use in numerous mine shafts in different underground mines all over the world. Besides its flaws, it does have multiple advantages, associated with its installation and operation, as well as economic benefits. Installation of elastic guidance is much quicker and cheaper than the assembly of stiff guides, similar to its maintenance and potential replacement. Rope guidance allows conveyances to travel with high speed, because of the smooth guidance, without shocks and collisions with the guides.

However, elastic guidance does have an essential disadvantage, which is a necessity for installation of the stiff guidance on the levels. It is crucial to ensure safety of the operation of the hoist and is also required by Polish law. Because of this requirement, elastic guidance of the conveyances is rarely used in Polish coal mines. It is caused by multilevel exploitation, which enforces one mine shaft with its hoist to operate on many levels, which in turn requires installation of the stiff guidance on each of these levels.

The biggest inconvenience caused by the stiff guidance on the mine levels is the process of the conveyance's entry onto the guides, because of the conveyance's lateral movements, intensified by conveyance's braking and turbulent air flow in the vicinity of the level. The necessity for conveyance's speed reduction decreases the level of effectiveness of the mine shaft and a hoist.

For the purpose of improving the operation of elastic guidance and to raise the effectiveness level of the mine shaft and the hoist, a retractable guidance system was introduced on the level 960 m in the Leon IV shaft in the Rydułtowy Coal Mine (currently ROW Colliery, Rydułtowy department). A patented retractable guidance system allows the conveyance to travel through the mine level with its full speed, if people or materials are transferred to a different mine level or in case of transport to the level 960 m, providing conveyance's stability on this level.

The problem with stiff guidance on the mine levels, described above, was neither examined nor solved, because the rope guidance of conveyances is rarely used in Polish coal mines, mostly because of its disadvantages listed above. The presented solution is innovative, because it utilizes a hydraulic cylinder to stabilize the conveyance in a very safe manner. Additionally, any major intervention in the conveyance's construction is not required, as it was in similar historic constructions. The construction of the retractable guidance system and the mine level allows installation of swinging bridges, which is important to provide safe and reliable transport of materials to the level 960 m.

The described solution is a pioneering approach in the Polish mining industry. It helped to raise the level of effectiveness of the Leon IV shaft in the ROW Colliery, Rydułtowy department, where it was installed. It might also help to raise the safety and effectiveness level in other mine shafts in different Polish underground mines.

This book covers numerous issues of elastic guidance of conveyances in mine shafts, especially behaviour of the conveyance guided by the ropes, as well as historic and modern solutions of rope guidance in underground mines in Poland and other countries all over the world. The idea of the retractable guidance system, operating on the level 960 m in the Leon IV shaft in the ROW Colliery, Rydułtowy department, has been presented in detail, as well as its elements, including its power supply and control of the hydraulic system. Numerous simulations, calculations and tests carried out in the shaft are also presented.

About the author

Paweł Kamiński was born on 7 May 1984 in Żory, Poland, into a family with mining traditions. In 2010, he graduated from AGH UST and became a research assistant. In 2018, he received his doctorate. He has been a lecturer at his alma mater since 2020. In 2018, he completed postgraduate studies in the field of mine hoisting systems. He has been the head of the design department of PBSz S.A. (part of JSW Group) since 2015. Over the years, he has held numerous internships in different mining companies in Poland, Germany, China, Russia and other countries. He has authored over 120 different publications, including eight patents and numerous different works for different bodies of the mining industry. Privately, he is a collector of historical mining literature.

Acknowledgments

The author wants to thank his colleagues from Shaft Sinking Company (PBSz S.A.), especially management of the company and employees of the design department.

This book would not have been written without people who mainly contributed to the retractable guidance system construction: Piotr Bulenda from ROW Coal Mine and PhD Marek Płachno from LABNIZ to whom the author wants to express gratitude.

Chapter 1

State of the art

1.1 LEON IV SHAFT

The Rydułtowy hard coal mine is one of the oldest Silesian collieries located in the Rybnik coal area in southern Poland (Adamczyk 2012). Its predecessor, named Charlotte, started production in 1806, and it was one of the biggest mines in Silesia at that time. Rydułtowy, as the first mine in the region, was equipped with a steam engine. In 1855, railway connected the mine with other regions of the country, increasing coal sales. In 1922, the area in which the Rydułtowy mine is located was incorporated to Poland. At the beginning of the 20th century, the colliery passed through numerous crises, which led to a reduction in employment and even the mine's closure for a four-year period, which started in 1932. The Rydułtowy coal mine was dynamically developing during the Second World War, because of the German economy's high demand for coal. In this period (1940–1944), employment increased threefold, up to 3,582 employees. After the Second World War, the mine was renamed Rydułtowy. Through the years, it belonged to various structures of the Polish mining industry (Jaros 1984). In 2004, the adjacent Rydułtowy and Anna collieries were joined and renamed Rydułtowy-Anna. Restructuring of the Polish mining industry led to merging of coal mines Jankowice, Marcel, Chwałowice and Rydułtowy, located in Rybnik and its vicinity, into the ROW colliery, belonging to the newly established Polska Grupa Górnicza (PGG, Polish Mining Group) (Wysocka-Siembiga 2016a).

A new shaft was sunk in the Rydułtowy colliery in the period of 1990–1998. It is called Leon IV, and its diameter is 8.5 m, which is rare in Polish mines. Shaft's depth in 1998 was 1,076.2 m, which enabled the mine to develop a new level on the depth of 1,050 m. It was considered that the 1,050 level covers the mine's need for coal.

However, rich resources on the depth over 1,000 m and the necessity of avoiding sub-level exploitation led to a decision of Leon IV shaft deepening to the depth of 1,210.7 m to develop another level on the depth of 1,150 m. Designing works for the purpose of shaft's deepening and extension of two operating hoists (main and auxiliary) started in 2013 (Wysocka-Siembiga 2016b).

The Leon IV shaft is one of the three mine shafts in Polish coal mines equipped with elastic (rope) guidance of conveyances. Thus shaft deepening and hoists extension to the depth of 1,150 m constituted a great challenge both for designers and people working in the shaft.

It should be noted that a mine shaft during its deepening has to be in continuous and undisturbed regular operation despite the work undertaken on its bottom. Such restriction

DOI: 10.1201/b22695-1

requires specific solutions. Actions taken for the purpose of the shaft deepening has to be carried out under the protection of specific construction. It might be either a layer of rocks or a construction called artificial shaft bottom. Its purpose is to separate two sections of the shaft and to protect people working on the shaft's bottom from falling objects.

However, in such cases, it is needed to provide means of vertical transport to the shaft's section below the artificial bottom. It is necessary to install hoists for transport of people and materials in the mine working located in the vicinity of the shaft under its artificial bottom.

A big facilitation in the process of shaft deepening is a pilot hole drilled in the shaft axis. Such borehole of diameter 500 mm or more is used for transport of excavated material, drainage and upcast ventilation of the shaft's face. However, utilization of the pilot hole requires a horizontal working on the destination level (Czaja and Kamiński 2017; Kamiński 2020).

The Leon IV shaft is an object in which many innovative solutions were applied. Some of them are presented below:

- single-layer waterproof shaft lining within the depth interval of 782.0−932.0 m,
- shaft deepening technology to the depth of 1,210.7 m,
- extending of shaft hoists from the level 1,000.0 (960 m) to the level 1,050 m and auxiliary level 1,200 m used for the mine drainage.

The initial project of the Leon IV shaft lining comprises two-layered lining with hydro-insulating shield made of PE foil on the section 782.0–932.0 m. It was necessary because of the occurrence of aggressive sulphate and magnesium water at this depth. Such lining construction was commonly used by KOPEX – Shaft Sinking Company (KOPEX-PBSz SA). Specific lining design consists of the following (Wichur, Frydrych and Kamiński 2015):

- B15 (C12/15) class concrete shaft set between 784.5 and 786.0 m;
- preliminary lining made of B15 (C12/15) class shaft panels between 786.0 and 932.0 m built from top to bottom;
- B30 (C25/30) class concrete shaft set between 930.5 and 932.0 m;
- Sealed and welded hydro-insulating layer of 2-mm-thick PE foil;
- Final shaft lining made of B25 and B30 (C20/25 and C25/30) class concrete built from bottom to top.

At the time Leon IV shaft was designed, one of the Polish cement producers made a special Portland cement, called the bridge Portland cement, marked with symbol CP 45(M), which was highly resistant to sulphate and magnesium corrosion. After the research executed in AGH University of Science and Technology, modification of the initial shaft lining construction and technology was proposed. It was suggested to make a 0.65-m-thick single-layer lining, built from top to bottom. Depending on foreseen pressure acting on the lining, its bearing capacity was modified by changing the concrete's composition and its class (Kostrz, Olszewski, Czaja, Deja, and Witosiński 2000). Two formulas were in use – the one marked with symbol R25/1/2 for B25 class concrete and R30/1/4 for B30 class concrete. The base of every concrete mixture was CP 45(M) bridge Portland cement (Kostrz et al. 2000).

All of the concrete receipts, developed in AGH University of Science and Technology and verified by laboratory tests, provided proper concrete strength and guaranteed suitable shaft lining's bearing capacity of designed thickness, as well as demanded water resistance of the level W8.

Concrete lining was made in 4-m-long sections in the direction from the top to the bottom. In case of using such technology, lining waterproofness depends primarily on technological joints between the upper (old) section and the bottom (new) section. In the descripted project, re-sealing of these joints was made using FUKO 2 injection hoses (Figure 1.1), for the first time ever in Polish shaft building. The application of this technology required an opinion of the State Mining Authority, concerning its security due to the presence of methane, after special tests conducted in Experimental Mine Barbara in Poland. The injection hoses FUKO 2 were attached to the upper section of the shaft with the use of metal connectors fitted with screwed joints. After another step of concreting, the gap was filled with a binding mixture on its whole length (Figures 1.1 and 1.2) obtaining satisfactory sealing of the key element of the shaft lining (Kostrz et al. 2000).

Another problem solved in the shaft building process comprised drainage of the rock mass behind the lining. It is commonly known that water swell behind the waterproof lining is dangerous because of the possibility of the occurrence of high hydrostatic pressures into the shaft lining, resulting in joining different water-bearing horizons by the shaft. This problem was solved by laying vertical drainage pipelines with a diameter of 100 mm along the shaft lining distributed on four azimuths (Figure 1.2).

In case of waterproof single-layered shaft lining used in the Leon IV shaft, the key issue is a waterproof concrete resistance to sulphate aggression. The concrete mixture was prepared in a professional concrete plant located at a distance of 40 km from the shaft. Thus, the concrete mixture had to be designed in such a manner that after bringing it to the shaft, it kept suitable consistence measured by the cone fall of about

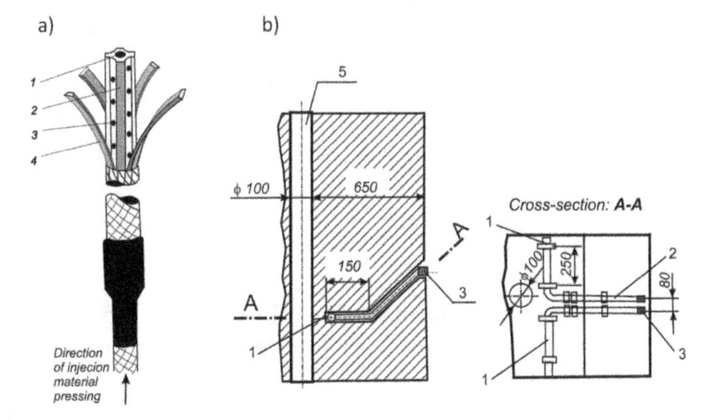

Figure 1.1 Sealing system of the concrete shaft lining with use of hoses FUKO 2. (a) FUKO 2 injection hose; 1 – injection channel Ø10 mm, 2 – hose core; 3 – injection holes; 4 – neoprene ribbons playing role of non-return valves. (b) Housing of hoses in technological joint of sequent sections of the concrete shaft lining; 1 – FUKO 2 injection hose, 2– steel pipes, 3 – threaded ending for pressure hoses, 4 – technological joint between two concrete lining section, 5 – drainage pipeline.

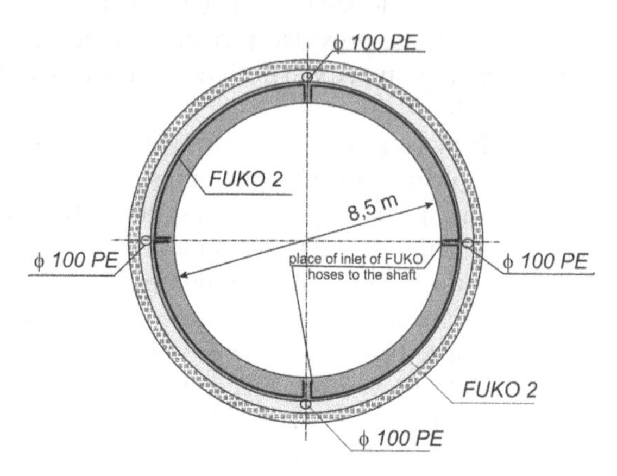

Figure 1.2 Injection system via hoses FUKO 2 and rock body drainage system behind the shaft lining.

12 cm. Additionally, admixtures of slag binder, plasticizer SK-1 and polypropylene fibre called Fibermesh have been introduced to the concrete mixture receipt.

The elaborated concrete mixture receipts based on sulphate-resistant bridge cements CP45(M), as well as constant author supervision of concreting works in the shaft resulted in success. Replacing the two-layered lining with waterproof single-layered lining allowed reduction of the cost of this shaft section by about 30%. This solution can be positively assessed after almost 20 years of exploitation, and it can be recommended to use in similar applications, particularly in case of aggressive water occurrence (Kostrz et al. 2000).

This investment of about 80 million PLN of the Leon IV shaft deepening by another 140 m resulted from the necessity of the modification of production processes in the Rydułtowy mine (Olszewski et al. 2017). This modification comprised shortening of personnel transport time to mining areas, as well as facilitation of the delivery of needed materials and considerable improvement of the ventilation of this part of the mine.

The investment task related to development of mine infrastructure in the Leon IV region comprised the following activities (Wowra, Nowak, Witkowski, Izydorczyk and Kamiński 2017):

- shaft deepening technical project,
- shaft deepening and equipment,
- the two-way inlet on the exploitation level at the depth of 1,150 m,
- the single-way inlet at the depth of 1,200 m designed for the needs of the main drainage system of the mine,
- elongation of mining hoists: main to the level of 1,150 and auxiliary to the depth of 1,200 m,
- installation of needed elements of mechanical equipment of the inlet on both built levels.

The target depth of 1,210.7 m was reached in August 2016 (Figure 1.3). After completion of the equipment of the deepened shaft section in 2017, works related to shaft

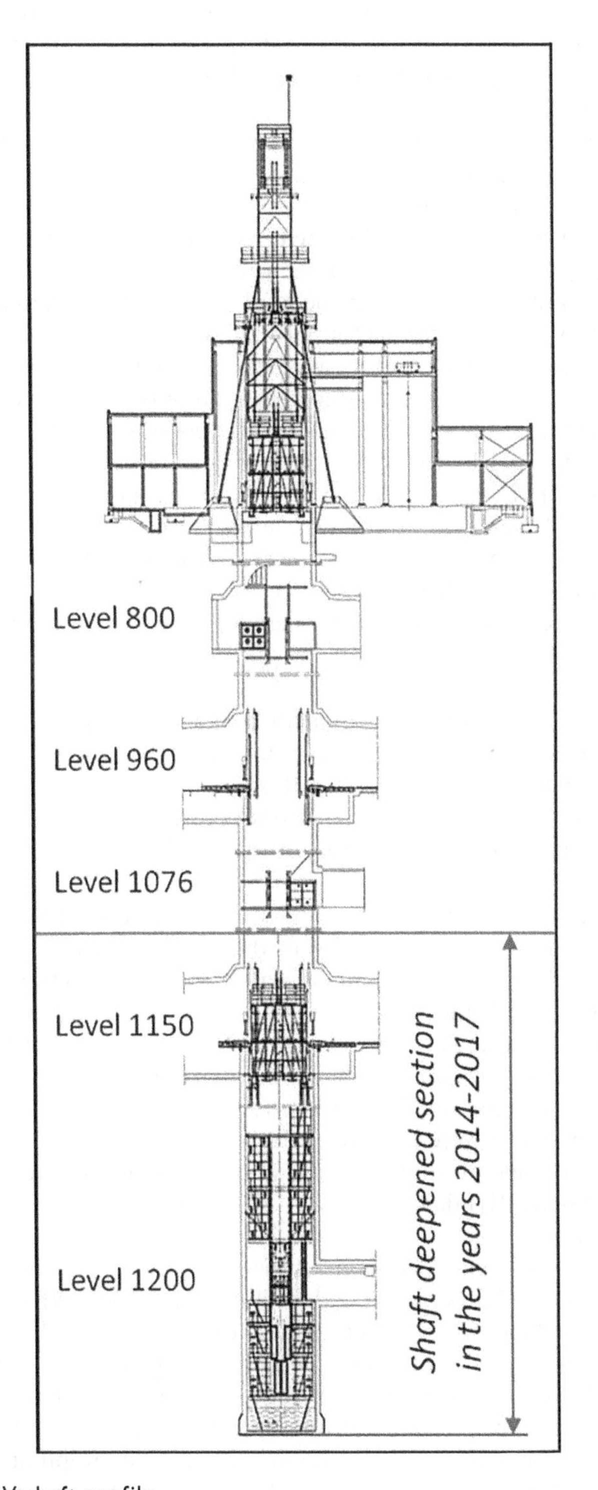

Figure 1.3 Leon IV shaft profile.

hoists' elongation began. The unusual and difficult task of guide ropes' extension was a pioneering venture in the Polish mining industry (Wowra et al. 2017).

Work related to deepening of the Leon IV shaft was conducted under the artificial shaft bottom, which was a construction comprising two decks connected with vertical partition. The application of such equipment was necessary to provide full operational ability of the shaft to the level of 960 m and the safety of staff working in shaft heading, beneath the artificial shaft bottom.

Mining work was conducted using the typical drill and blast method. As winning haulage is the biggest issue in the shaft deepening process, a horizontal drift was excavated in the vicinity of the shaft at the level 1,200. A large-diameter borehole was drilled from this working to connect it with the existing shaft bottom. The borehole was located in such a manner that its axis was situated 2.2 m towards east from the shaft axis to allow safe and collision-free operation of the cage covering the borehole. However, the 1,200-mm-diameter borehole was exposed to the danger of development of the winning jam. In order to remove the jam in the hole, a rope of 25 mm diameter with conveyor scrappers was installed in the hole. The rope was set into motion by two KUBA-5 windlasses, installed in drifts on levels 1,076 and 1,200 m (Figure 1.4). The movement of conveyor scrappers allowed to break rocks blocking the borehole, resulting in borehole unblocking (Kamiński 2020).

Transport in the whole deepening process was handled by special devices located in a drift on level 1,076 m (Figure 1.4):

- B-1500/Ex/AC-2m/s hoisting machine for bucket handling,
- two KUBA-10 low-speed windlasses for formwork handling,
- two KUBA-5 windlasses for rope handling,
- KCH-9 windlass for manoeuvring of basket protecting the borehole,
- supporting construction for assembly of sheave wheels.

In difficult conditions in Leon IV shaft, limited space in the drift on the level 1,076 and explosion hazard, innovative construction of the hoisting machine was necessary. The B-1500/Ex/AC-2m/s hoisting machine was designed and produced by the Polish company MWM Elektro, specifically for the purpose of Leon IV shaft deepening. The model of the hoisting machine is presented in Figure 1.5 (Kowal 2013; MWM advertising materials).

The B-1500/Ex/AC-2m/s hoisting machine was designed for installation in underground workings, including methane and coal dust explosion hazard zones. The dimensions of the machine were carefully selected to enable the transport of modules and components in underground workings and installation in the desired operation location near the shaft (Ryndak and Kowal 2015; Madej, Radowski, Kowal, Turewicz and Helmrich 2013).

The B-1500/Ex/AC-2m/s hoisting machine is used to drive single-end mine shaft hoists in major, minor shafts and sloped workings. The B-1500/Ex/AC-2m/s hoisting machine was granted a permanent approval by the President of the State Mining Authority on 18 December 2015. The machine can be controlled manually from the control panel installed in the ergonomic cabin, protecting against noise and dust or can be controlled remotely from inspection and operation stations, in remote travel mode with smooth speed adjustment. Figure 1.6 presents a photograph of the hoisting machine and Figure 1.7 displays its diagram (MWM advertising materials).

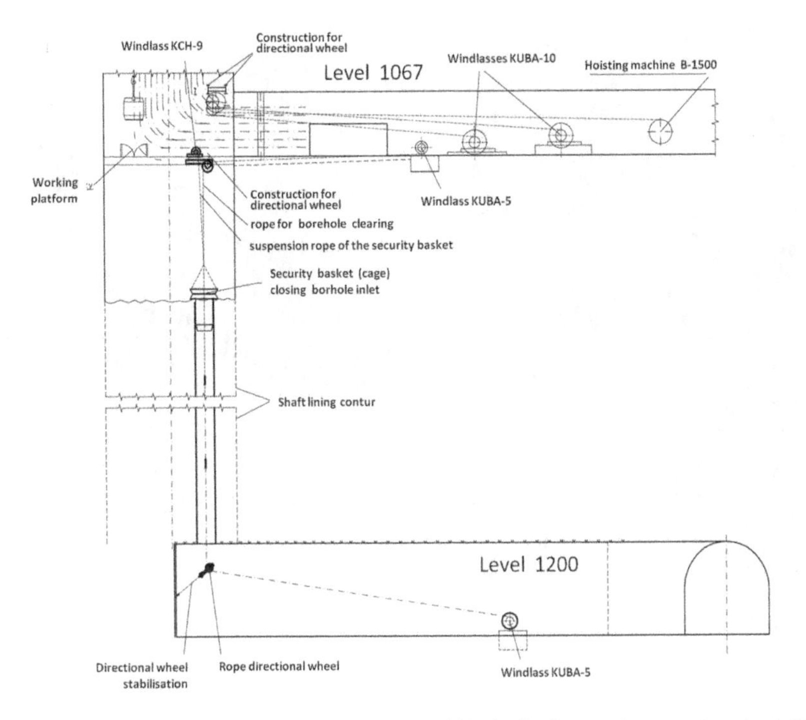

Figure 1.4 Distribution of devices during Leon IV shaft deepening on levels 1,076 and 1,200 m.

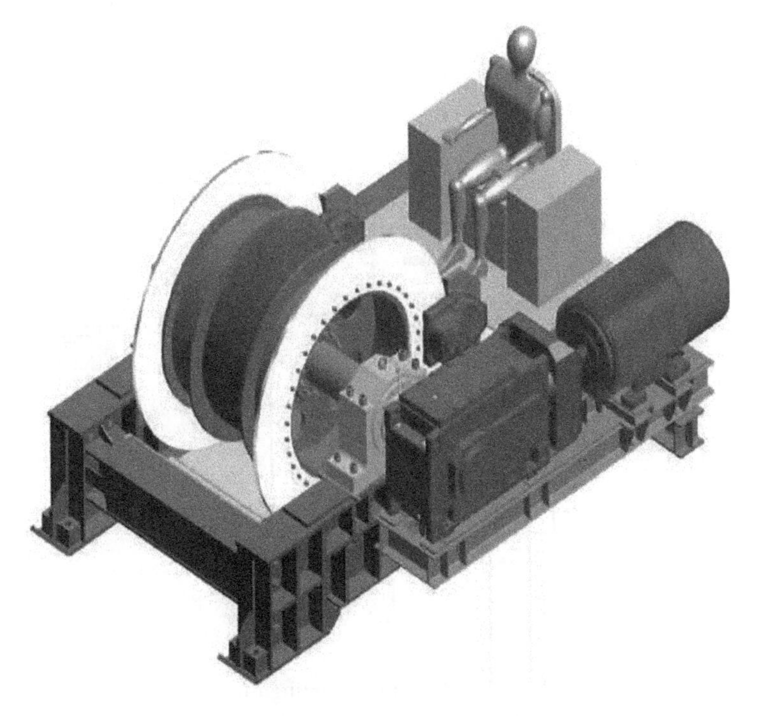

Figure 1.5 Model of the B-1500/Ex/AC-2m/s hoisting machine.

Figure 1.6 A view of the B-1500/Ex/AC-2m/s hoisting machine.

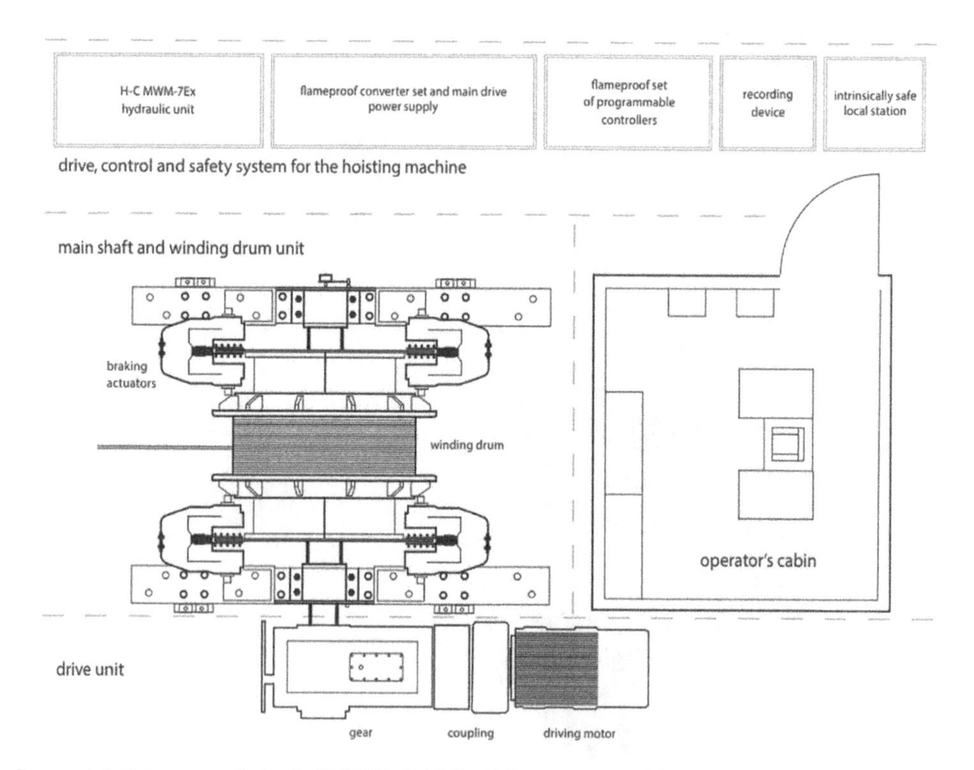

Figure 1.7 Diagram of the B-1500/Ex/AC-2m/s hoisting machine.

The main components of the mechanical part are the following:

- main shaft unit,
- drive unit with asynchronous motor and gear,
- hydraulically controlled brake.

The main components of the electrical part are the following:

- power supply and drive unit,
- control and adjustment system,
- safety system,
- visualization and recording system

The main shaft unit of the hoisting machine consists of a winding drum (hoist drum) with the diameter of 1,500 mm mounted on a shaft supported by two rolling bearings: floating and retainer. The drum has a LeBus-type lining, guaranteeing (in case of multi-layer winding) the correct laying of rope on the drum and even pressure on the coat surface. The winding drum is fitted with two side brake discs, interfacing with the braking system actuators (Madej et al. 2013; MWM advertising materials).

The drive unit of the machine consists of a gearbox mounted on the main shaft (ratio = 56) and a three-phase asynchronous motor (power = 132 kW). The motor is connected to the high-speed shaft of the gear via a flexible coupling. The drive motor is powered and controlled via a frequency converter in an explosion-proof housing. Due to the implemented software, the set drive unit parameters can be maintained with stability. The application of this type of power supply does not negatively affect the power grid – adverse harmonic currents and voltages have been reduced to minimum (Madej et al. 2013; MWM advertising materials).

The braking system consists of the following:

- four braking props, each with one pair of brake actuators (BSFG 405), interfacing with two brake discs,
- hydraulic system for brake actuators and connecting system elements,
- electrohydraulic control and supply unit type H-C MWM-7Ex,
- emergency oil discharge device.

The control, regulation and safety systems are based on a Flameproof Set of Programmable Controllers. They are based on the implemented digital travel controller GRZ-13-A (in the single-end hoist version). Its task is to correctly calculate the current position and speed of the conveyance and create the diagram of travel in the function of the path, based on the calculation (Madej et al. 2013; MWM advertising materials).

All the important functions describing the operating conditions of the hoisting machine/mine hoist are included in the screen visualization system, based on a flameproof computer and a visualization application by MWM Elektro Sp. z o.o. The application provides the user with full control of the operation of the hoisting machines, increasing the efficiency and safety of the operation of the mine hoist (Madej et al. 2013; MWM advertising materials).

Parameters of the B-1500/Ex/AC-2m/s hoisting machine are presented in Table 1.1. In Figure 1.8, a view of the hoisting machine in the drift on the level 1,076 during works for Leon IV shaft deepening is presented.

Single-layer C30/37 concrete class lining was designed for the new section of the Leon IV shaft. Concreting was conducted using a 2.15-m-tall sliding formwork. Lining, calculated and consulted with mine services, has a thickness between 0.5 and 0.6m. Due to small and temporary forecasted water inflows into the shaft, no additional water protection was designed.

Thanks to described investment, Ruch Rydułtowy of ROW colliery was allowed to exploit coal from seams no. 713/1-2 and 712/1-2 in a safe and effective manner. These coal seams belong to the most promising mining assets within mining areas belonging to this part of Rybnik Mining District.

The development of this part of the deposit allowed establishing a new level at the depth of 1,150m. The two-way inlet is equipped with a full set of the wheel transport handling, with special platform for material re-loading from wheel into the suspended transport system (Kamiński 2020).

Table 1.1 Parameters of B-1500/Ex/AC-2m/s hoisting machine (Ryndak and Kowal 2015)

Machine Type	*B-1500/Ex/AC-2m/s*
Rated power	132 kW
Rotational speed	1,480 rpm
Drum diameter	1,500 mm
Maximum static force in the wire rope	50 kN
Maximum lifting rope breaking force	440 kN
Maximum travel speed	2 m/s
Acceleration/deceleration	0.4 m/s^2
Total weight	~15,850 kg
Recommended distance to a sheave wheel	min. 15 m; max. 30 m

Figure 1.8 A view of the B-1500/Ex/AC-2m/s hoisting machine in the drift on the level 1,076.

The inlet parameters are as follows:

- excavation founding depth – 1,143.7 m,
- height – 7.3 m,
- width: W side – 8.11 m, E side – 7.0 m,
- basement depth – 2.30 m.

Cross section of the inlet is shown in Figure 1.9, and its 3D model is shown in Figure 1.10.

Universality is a characteristic feature of the shaft inlet at the depth of 1,150 m – the main transport level. On this level, there is a possibility of using three shaft hoists,

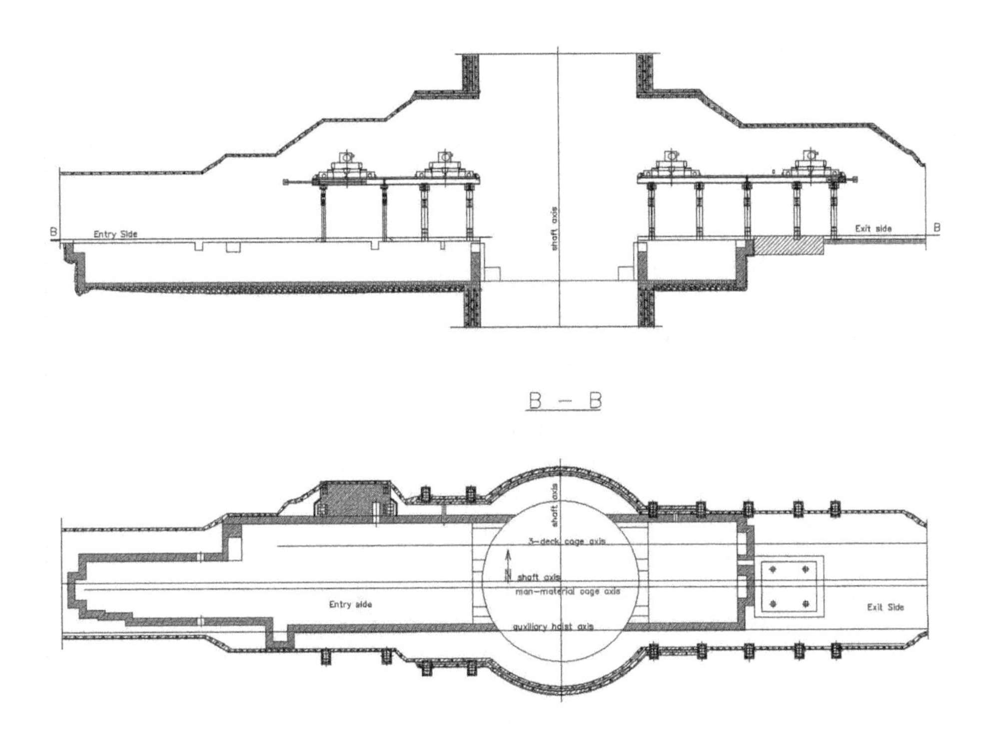

Figure 1.9 Inlet on the level 1,150 m.

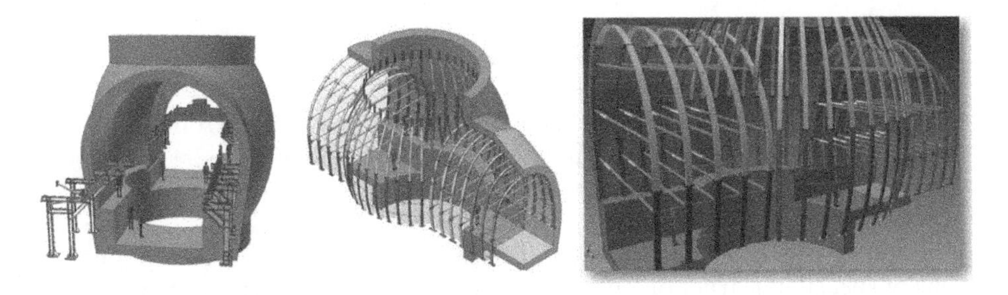

Figure 1.10 3D model of the inlet on the level 1,150 m.

which might considerably accelerate the process of material transport, as well as it will allow fast and fluent personnel transport. Using transport platforms in the mine transport system forced equipping shaft station basement with devices and machines needed for pushing mine carts into man-material cage, as well as into other conveyance. During Leon IV shaft deepening, water management in this region was ordered. For this purpose, excavations needed for the main water drainage handling were localized in the shaft station of the level 1,200 m. Also, elongation of the auxiliary host to this level was necessary. It was also re-qualified from auxiliary hoist into the so-called "small" hoist (Wowra et al. 2017; Kamiński 2020).

Geometry of the shaft inlet on level 1,200 m is as follows:

- depth of the excavation founding – 1,195.7 m,
- height – 4.2 m,
- width – 6.1 m.

The shaft inlet is supported with anchor-concrete-steel lining and is equipped with shaft chairing construction with swinging bridges in the inlet basement.

As the main transport shaft, the Leon IV shaft is equipped with two compartments: one for main hoist with man-material cage and another for standard three-deck cage and auxiliary hoist. The shaft cages are suspended on two 48-mm-diameter rope carriers driven by Koeppe drive wheel hoisting machine. In order to balance masses of rope carriers, two balance ropes of diameter Ø53 mm are installed (Kamiński 2018).

Conveyance's guidance in the Leon IV shaft is realized using rope guidance. Guide and rubbing ropes are suspended using glands on the head frame and tensioned using cheese wheels of mass, which provides tension of at least 8 kN/100 m of the rope. These cheese weights are assembled in the shaft sump in cases of special construction. Elements of the guidance and other shaft equipment are shown in Figure 1.11 (Nowak, Bulenda, Piszczan and Kamiński 2019; Kamiński 2020).

Elastic guidance of the shaft conveyances comprises 12 guide ropes and four rubbing ropes between man-material and three-deck cage. Three rope carriers and balance ropes are also used, one for each conveyance. The layout of ropes used for conveyances' guidance is shown in Figure 1.12 (Czaja, Kamiński, Olszewski and Bulenda 2018; Kamiński 2020).

The other element of the shaft furniture is related with main hoist cage braking system localized under the level 1,150 m and in the head frame. This system consists of thickened wood guides. The other elements of the shaft equipment are the following (Czaja et al. 2018; Nowak et al. 2019; Kamiński 2020):

- platform of the return station of balance rope,
- platform of balance rope control,
- positioning frame of cheese weights of guide ropes and rubbing ropes,
- control platforms of guide and rubbing ropes and cheese weights,
- safety platform.

Application of elastic conveyance guidance in a mine shaft requires using chairing systems of special construction on the levels, equipped with co-called angular guides, which are basically stiff guides assembled in the vicinity of a level. In case of the Leon

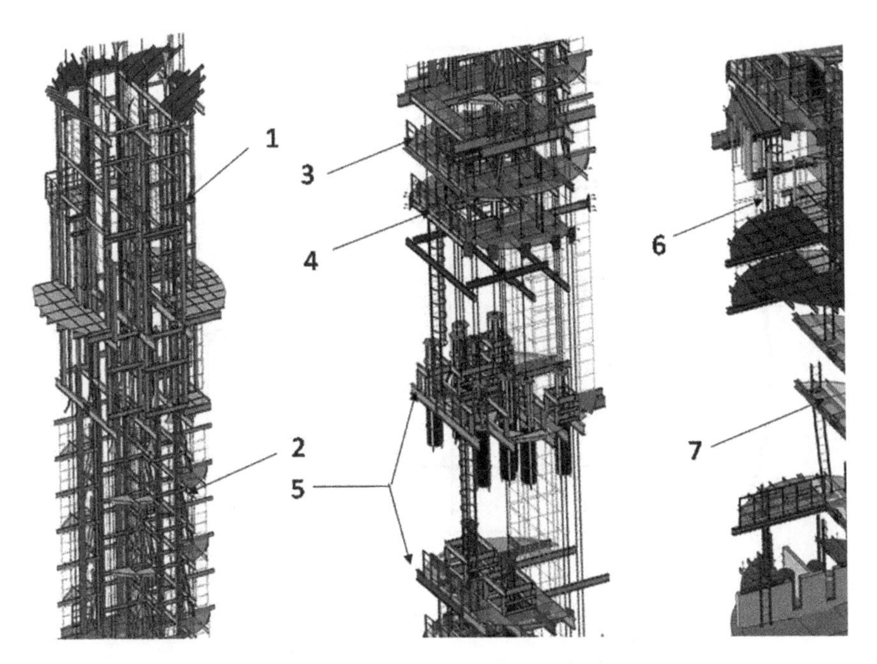

Figure 1.11 Model of the shaft Leon IV furniture after deepening to the depth of 1,210 m. 1– level guidance on level 1,150 m; 2 – braking system of the main shaft hoist; 3 – return station of the equalizing ropes; control platform of the equalizing ropes; 5 – control platforms of guiding rope weights; 6 – level guidance on level 1,200 m; 7 – furniture of the shaft Leon IV– sump.

IV shaft, application of this construction was necessary on levels 800, 1,076, 1,150 and 1,200 m (Czaja et al. 2018; Nowak et al. 2019).

Man, mining or material shaft deepening always requires elongation of the hoisting system used in the shaft. In case of the Leon IV shaft, assembly of the hoisting system in the process of shaft sinking comprised using ropes significantly longer than necessary. Excess ropes were stored on drums of special construction at the shaft station on the level 1,076 m. Thus, elongation of guides required only control of ropes wear by experts and lowering them to the level 1,150 m (Olszewski, Czaja, Bulenda and Kamiński 2018).

Work related to hoisting system elongation began with auxiliary hoist rope guides in number of two Ø32 mm ropes, which were lowered to the level 1,200 m. The weight of the rope of diameter Ø32 mm is about 5.3 tonnes, while for Ø54 mm rope, it exceeds 25 tonnes. Every rope manoeuvre, such as lifting or lowering, requires application of special winch with high lifting capacity. This winch has to be assembled on the bank level, and it is also necessary to use sheave wheel on top of the head frame. For the purpose of moving ropes in the shaft, EWP-35 hoist was used. After making welded clamps and taking rope weight by the winch, glands were disassembled, and ropes were moved towards the surface level to be inspected by an expert. After positive decision of the expert and placing ropes into service, they were lowered down to the level 1,200 m, where cheese weights were attached (Czaja et al. 2018; Olszewski et al. 2018; Nowak et al. 2019; Kamiński 2018, 2020).

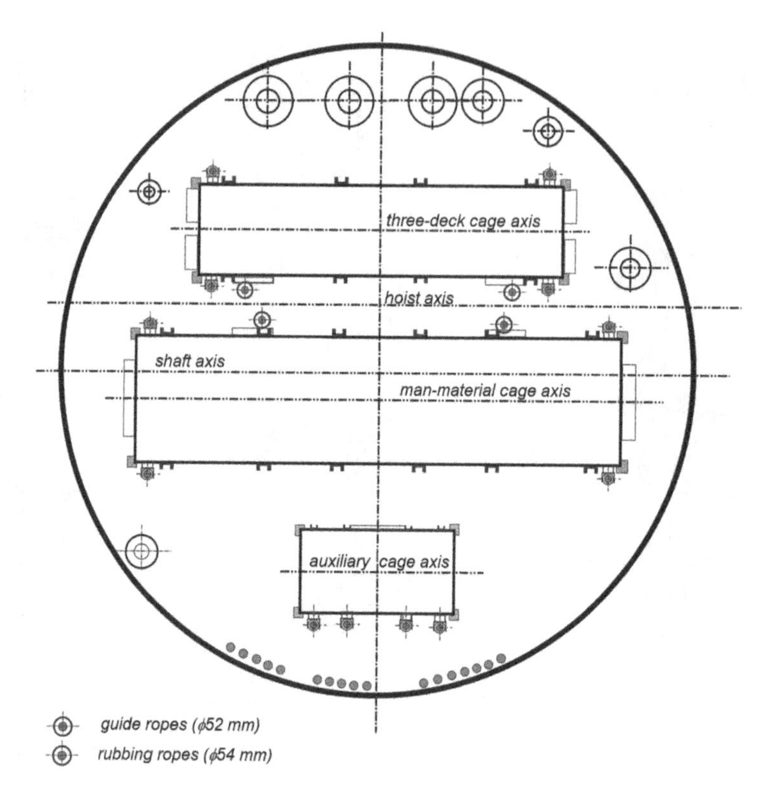

Figure 1.12 Elements of the elastic guiding of shaft Leon IV.

Another stage of hoisting system extension was elongation of rubbing ropes and guide ropes of the main hoist, which was to reel out the rope excess, stored at the level 1,076 m. After assembly of additional sheave wheel construction, which is needed to guide technological Ø40 rope in the area of the gland, on the bottom sheave wheel platform, main rope extension works began. After making welded clamps and taking rope weight by the winch (40 tonnes), ropes were moved towards the surface level and then to the level of the control platform. After the process for a rope was completed, sheave wheel was moved into another position. This operation was repeated eight times for guide ropes and four times for rubbing ropes, assembled between conveyances. Figure 1.13 presents the diagram of equipment necessary for the process of hoisting systems' extension and model of sheave wheels' attachment to the head frame (Kamiński 2018).

The technology of the hoist and balance ropes' replacement in the Leon IV shaft is similar to the standard procedure of periodic rope replacement in operating mine shafts. At first, preparatory work was done, comprising of building a base for EPR-1000 friction winch and assembly of sheave wheels at the bank level. Hoist ropes were lowered down in the shaft after placing conveyances at the platform and taking the weight of ropes by the friction winch.

After moving the man-material cage to the level 1,150 m and assembly of suspensions, balance ropes were extended. In this case, man-material cage, moving upwards, pulled new balance ropes. After moving the three-deck cage to the level 1,150 m,

Figure 1.13 Visualization of sheave wheel location on hoist tower (Kamiński 2018).

installation of new balance ropes under it was conducted (Kamiński 2018; Bojarski, Bulenda, Kamiński and Nowak 2019)

1.2 ROPE GUIDANCE OF CONVEYANCES

Rope guidance of conveyances has been in common use for decades in numerous mine shafts all over the world. It was especially popular in underground mines in the UK and Republic of South Africa (RSA), where it is still widely seen. In the mines of continental Europe, stiff guidance of conveyances was more popular than the elastic one. However, rope guidance was also used in some of the German, French or Russian mines (Slonina and Steuhler 1980). Nowadays, elastic guidance of conveyances is still used in many shafts, inter alia in Chinese mines.

Multiple variants of elastic guidance are used in mine shafts. They differ according to the location of the conveyance regarding the shaft's cross-section, number of shaft's compartments, shaft's diameter or function, etc. (Slonina and Steuhler 2020; Olszyna et al. 2018; Tobys and Tytko 2011).

Elastic guidance of conveyances requires specific parameters of ropes to be used as a rope guide. Theoretically, any type of the rope can be used as a guide or rubbing rope; however, not every type performs this function properly. The main requirement for the guide or rubbing rope is its high transverse stiffness. Such rope has to be

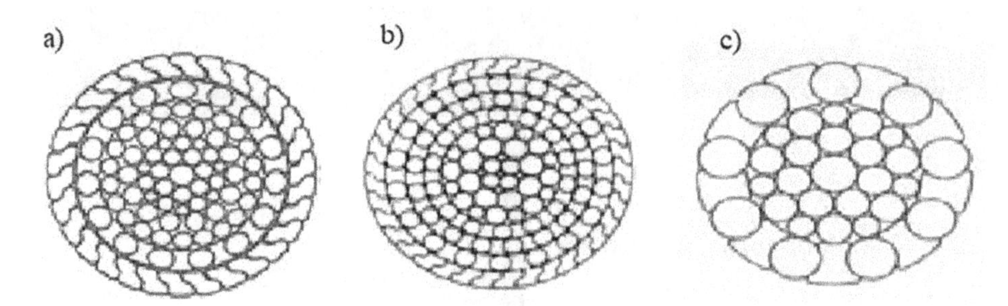

Figure 1.14 Cross sections of ropes used as guide and rubbing rope. (a), (b) locked coil ropes and (c) half-locked coiled rope.

characterized with adequate strength and high resistance for outer wear in type of the mass loss. Of the many types of ropes available on the market, locked and half-locked coiled ropes work best (Slonina and Steuhler 1980, Mańka, Słomion and Matuszewski 2018; Olszyna et al. 2018; Tobys and Tytko 2011). Cross sections of ropes of these types are presented in Figure 1.14. Polish law forbids application of other types of ropes as guide and rubbing ropes (rozporządzenie 2016).

Guide ropes in mine shafts are installed on the head frame using suspensions. The most popular type of such suspension is a gland. Examples of glands are shown in Figures 1.15 and 1.16. For the purpose of rotation compensation, they are usually spherically seated, which allows insignificant deviation of the rope from the vertical (Carbogno and Żołnierz 2009; Tobys and Tytko 2011; Olszyna et al. 2018).

Less frequently used solutions of rope guides' suspension are conical-shaped glands with separator wedge, thimble-type glands with flat clamp (used in shallow shaft, which makes them vanishing) and wedge-crane glands (used in underground mines in Russia and Ukraine). Rope's straightness is provided by weights, attached to ropes with suspensions or stabilizing sleeves (Carbogno and Żołnierz 2009; Tobys and Tytko 2011; Olszyna et al. 2018). Figure 1.17 presents a typical type of gland used in Polish mines.

Proper mounting of rope guides is a very important issue in terms of safety of the hoisting system. In bibliography, examples of movement of the guides in the glands or even their slippage can be found. Safety of the guides' assembly in the glands depends on the following (Carbogno and Żołnierz 2009):

- wedge concurrence,
- state of wear of wedges and casing surface,
- friction coefficient between wedges and a rope and between wedges and casing,
- assembly of wedge clamps,
- initial pressure on wedges,
- cleanliness of rope and wedges,
- lubrication of wedges' surface at the contact with casing,
- rope hold certainty way of calculation.

Guide ropes are usually tensioned by weight loading in bottom rope ends. The most popular solution of tensioning is application of mass blocks, the so-called cheese

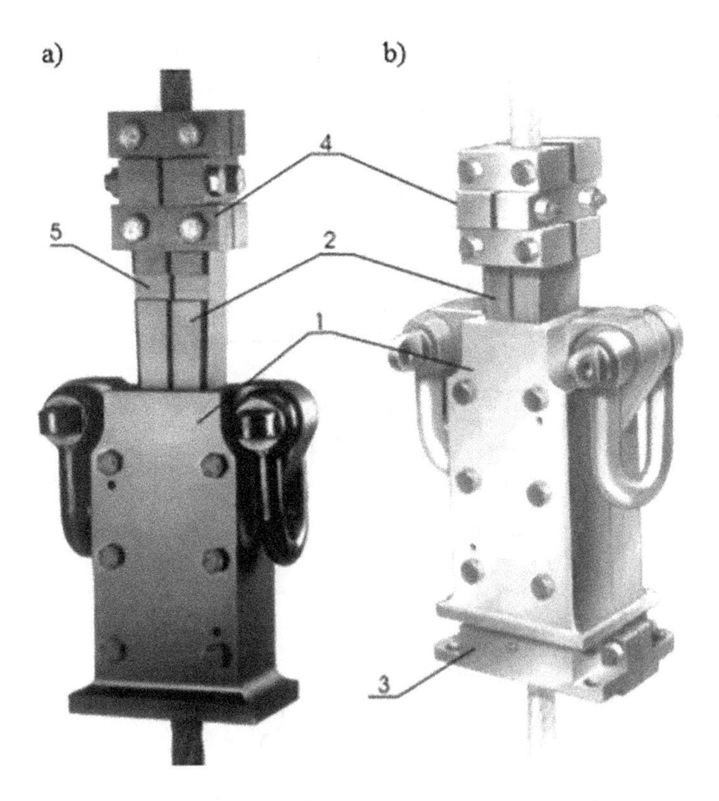

Figure 1.15 A view of glands made by Bellambie (Carbogno and Żołnierz 2009). (a) Regular gland and (b) articulated gland. 1 – casing, 2 – wedge, 3 – base with the spherical seat, 4 – screw terminal, 5 – excision for assembling of the tool for pushing or pulling out the wedge.

weights, which are basically cast-iron or leaden plates. An example of such construction is shown in Figure 1.18. Such solution allows eliminating the influence of ropes' elongation caused by temperature change. An alternative for cheese weights is spring or hydraulic loading. However, it is less frequently used than weight loading, and in Polish mines, it is not used at all. It is caused by the necessity of assembly of these constructions at the top end of the rope, which effects in high gravitational forces in the construction. The application of these systems at the bottom end of the rope is impossible because of the extremely hard environment in the shaft sump (salty water and humid air) and a necessity for power supply, which makes such solution impractical (Olszyna et al. 2018). Figure 1.19 presents a typical solution of guide ropes in Polish mines.

It is a rule of thumb that the value of the force for rope tensioning is 10 kN/100 m of the rope (Slonina and Steuhler 1980). Existing Polish legislation is even more strict, because in case of conveyances with one hoist rope and mass greater than or equal to 20 Mg, the required force is equal to 16 kN/100 m of the shaft depth. The force required by the law is also in accordance with the shaft depth (rozporządzenie 2016). Figure 1.20 presents a rope tensioning system for use in mine shafts.

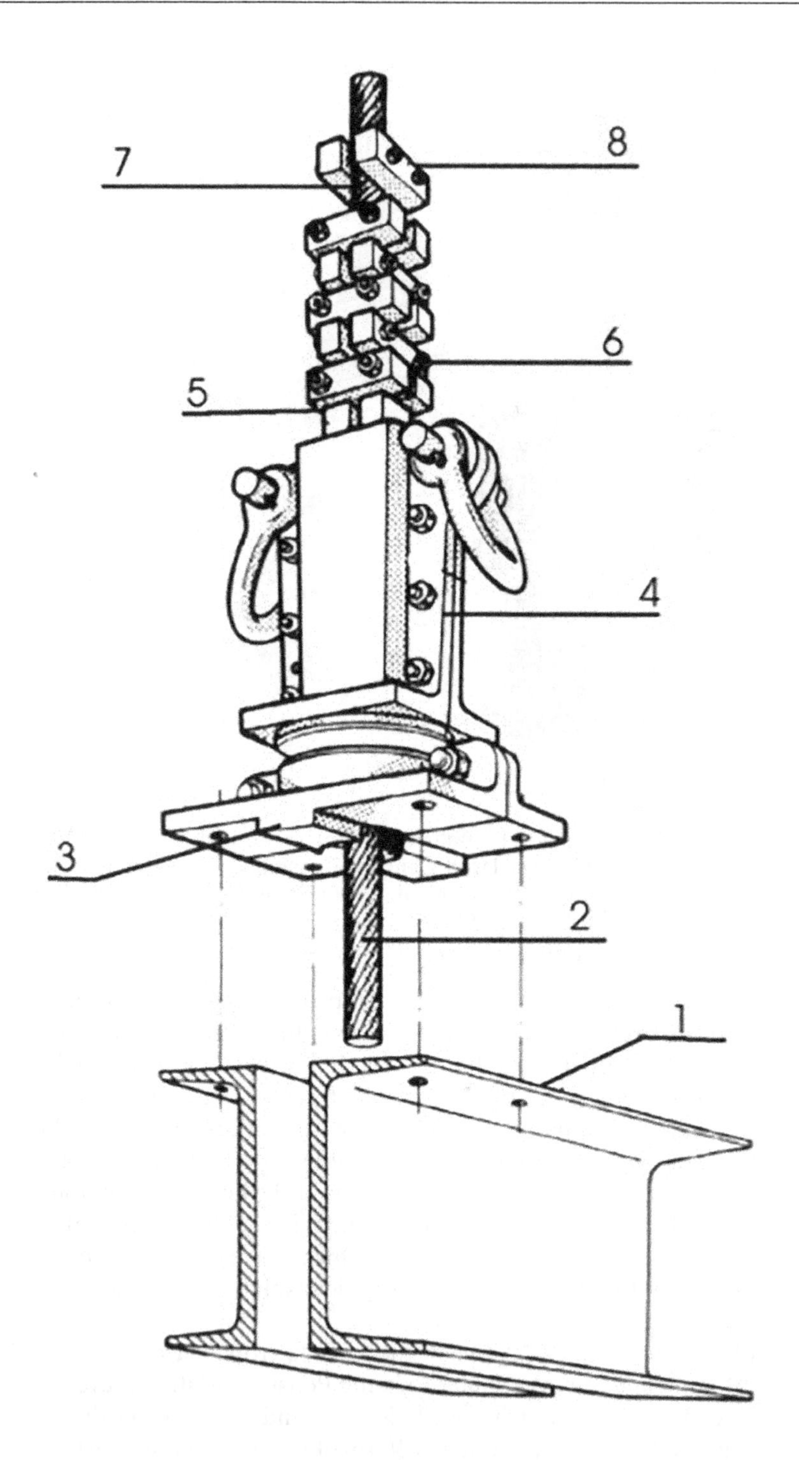

Figure 1.16 Assembly of the spherically seated gland (Carbogno and Żołnierz 2009). 1 – HEB beam, 2 – rope guide, 3 – base, 4 – casing, 5 – wedge, 6 – screw terminal, 7 – gap for rope control, 8 – indicator of the rope's movement.

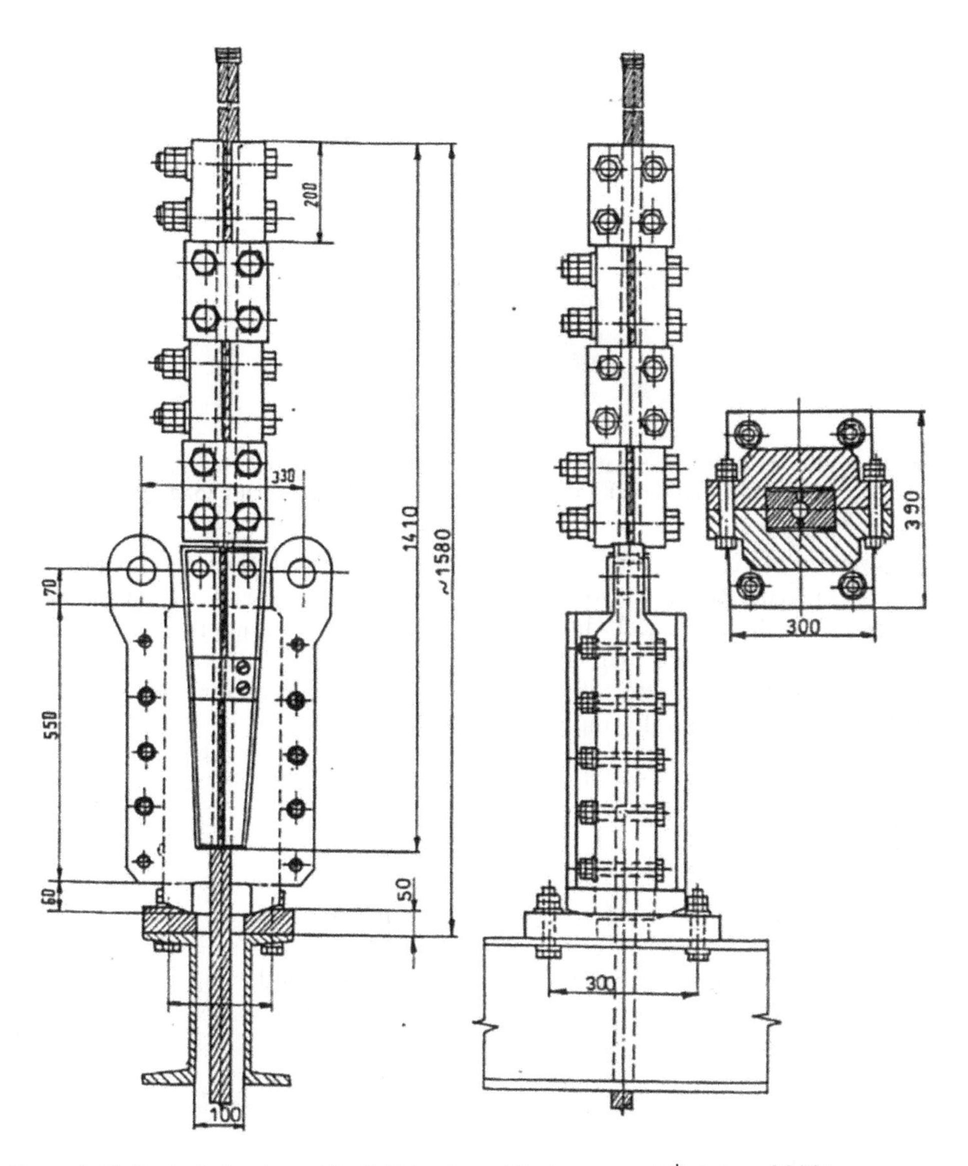

Figure 1.17 Typical gland used in Polish mines (Carbogno and Żołnierz 2009).

Different arrangements of guide and rub ropes are used in practice. They are classified into two groups – with all guide ropes placed on the side of the conveyance (along its long side) or with guide ropes located in the corners of the conveyance. Opinions on different arrangements expressed in the literature are diverse. Some of them prove superiority of the ropes' arrangement on the side of the conveyance, because of smoother run of the conveyance. However, this opinion is hard to verify (Slonina and Steuhler 1980). On the other hand, some authors favour corner arrangement because of its greater restoring moment (Bura 1970c).

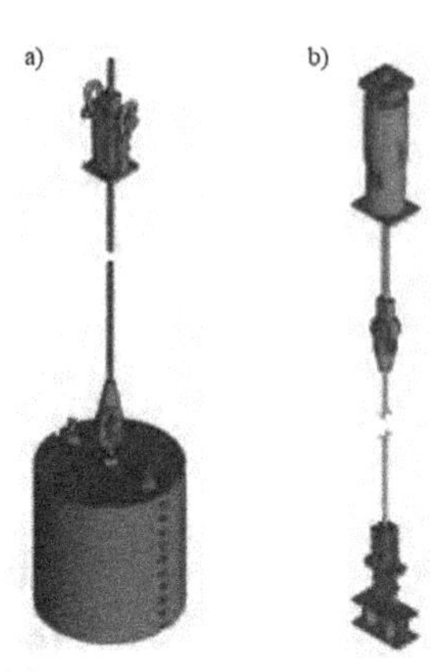

Figure 1.18 View of different tensioning systems by Northern Strands (Northern Strands advertising materials). (a) by weight loading, (b) by spring loading.

In some cases, rub ropes are used in mine shafts. Identical construction of rub and guide ropes is required. It is favourable to use rubbing ropes with greater diameters than guide ropes; however, according to practical problems, it is not required (Slonina, Steuhler 1980; Tobys, Tytko 2011; Olszyna et al. 2018).

There are numerous profits of rope guidance application in mine shafts (Olszyna et al. 2018), including the following:

- low cost of materials,
- easy handling,
- high durability,
- short assembling time,
- smooth run of the conveyance without shocks and side hits,
- high speed of the conveyance,
- higher hoist rope durability caused by conveyance's smooth run,
- ventilation resistances are almost ten times lower than in case of shafts equipped with stiff guides.

There are however disadvantages of rope guidance, which are as follows:

- greater technological gap between a conveyance and shaft lining and equipment caused by lateral moves of the conveyance, perpendicular to shaft axis,
- difficulty of operation caused by rock mass movement,
- requirement for greater diameters of the shaft and problems with locating shaft equipment in numerous sections,
- forces up to 2,000 kN required for rope tensioning loading head frame might force application of special strengthened head frame construction,
- rope tensioning system requires suitable shaft sump adaption,
- in case of using two conveyances, application of rub ropes is necessary to prevent the collision of conveyances moving in opposite directions,
- hoist rope has to be non-rotating rope.

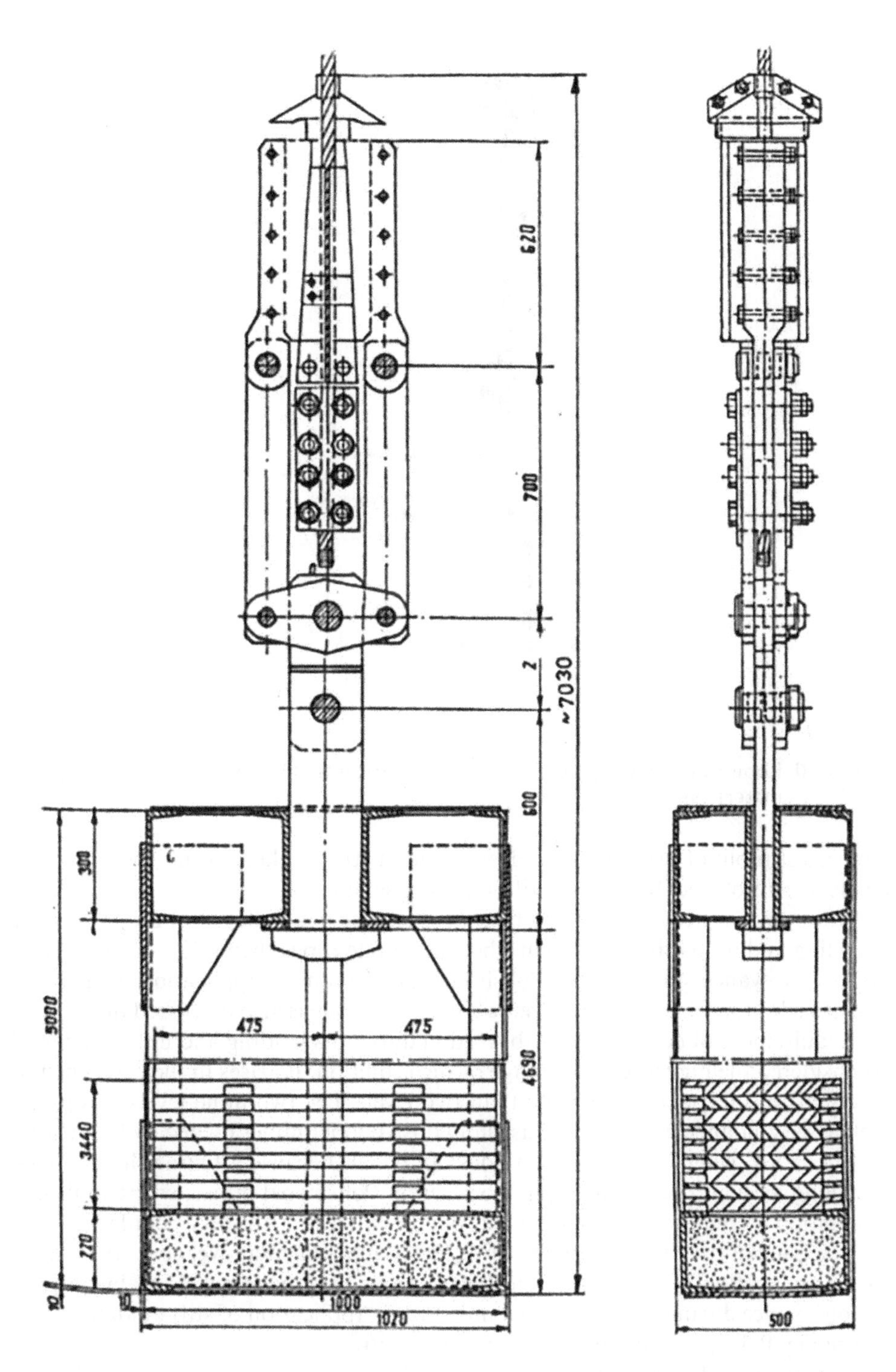

Figure 1.19 Typical way of rope tensioning used in Polish mines (Carbogno and Żołnierz 2009).

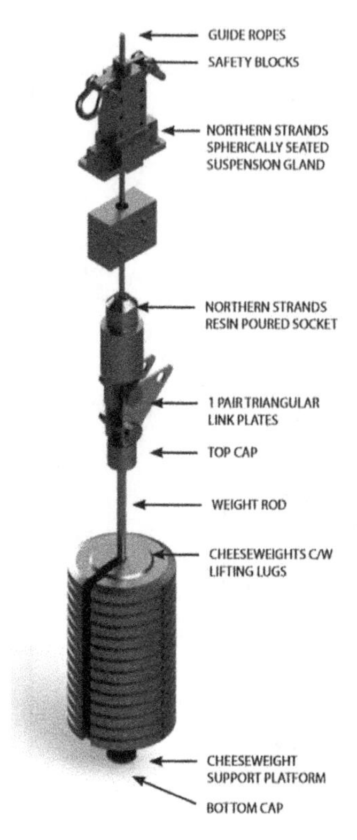

Figure 1.20 Rope tensioning system by Northern Strands (Northern Strands advertising materials).

The biggest profit of using rope guides instead of stiff guidance is no necessity to assemble dozens of steel frames for stiff guides' assembly.

However, a conveyance guided by rope guidance has to be stabilized for the time of loading and unloading operations, because guide ropes themselves do not provide proper conveyance stabilization. Polish legislation requires application of stiff guidance in such situations (rozporządzenie 2016). This issue is also discussed in this work.

Requirement of conveyance stabilization during its loading and unloading might be considered the biggest disadvantage of such solution. It arises in the case when one mine shaft has to operate on more than two levels, i.e., two mine levels and ground level. Rope guidance is commonly used in Polish copper mines owned by KGHM Polska Miedź SA, where such situation does not occur, because all mine shafts operate on ground level and one mine level. However, in Polish coal mines, where multi-level exploitation is common, elastic guidance of conveyances is not very popular.

The issue of operation on many mine levels, mentioned above, concerns a necessity for application of stiff guides on shaft stations to provide proper stabilization of the conveyance during its loading and unloading. Application of stiff guidance is also required by Polish legislation (rozporządzenie 2016).

The process of conveyance's entry on stiff guides and its ride through the shaft chair construction require reduction of conveyance's velocity to provide proper level of safety of the shaft and hoist. However, it negatively affects the effectiveness of the hoisting system. The necessity for speed reduction is caused by the importance of precise entry of the conveyance into stiff guide construction. In case of the Leon IV shaft, the velocity of the conveyance had to be reduced from 10 to 0.5 m/s. Moreover, speed had to be reduced twice during one ride cycle, because both conveyances had to go through the stiff guides, which are located at the level and which is not in the area of conveyances' passing. The tendency to increase the speed of conveyance in the vicinity of levels is natural and requires application of customized solutions. Construction used in the Leon IV shaft is pioneering in conditions of the Polish mining industry.

However, rope guidance of conveyances is commonly used in mine shafts of underground mines all over the world, and it was also used in Polish coal mines back in the days. The following sections present an overview of research carried out on behaviour of conveyances guided with ropes as well as historic and modern applications of rope guidance in Poland and other countries.

Rope guidance of conveyances has been successfully applied in mine shafts for decades. However, the behaviour of the conveyance guided by ropes has not been fully understood and explained for a very long time (Renyuan, Zhu, Chen, Cao and Li 2014). The key issue of this problem is to characterize movements of the conveyance perpendicular to the shaft axis, which might be used to determine required gaps between moving conveyances and shaft lining or equipment. Technological gaps were actually a rule of thumb, specified on the basis of experiences in rope guidance operation.

The issue of rope guidance in mine shafts of Polish underground mines was thoroughly examined by Lesław Bura, mostly in the 1960s and 1970s. His works are characterized by a complex and innovative approach, which is noticeable in the application of video cameras for analysis of rope-guided conveyance's movement (Bura 1971). Bura investigated almost every stage of operation of shaft and hoisting system, from its design and rope arrangement (Bura 1970c, 1972), rope guidance performance during shaft operation (Bura 1970a, 1975a) to shaft closure and working conditions of the elastic guidance during liquidation of the shaft pillar (Bura 1975b). Unfortunately, nobody decided to continue his work; thus, the majority of Polish work on rope guidance of conveyances is dated to the 1970s. More attention was paid to stiff guidance of conveyances because of their popularity in Polish mines.

The vast majority of research and analysis carried out by GIG Research Institute and described by Lesław Bura in numerous publications concern the issue of lateral movements of the conveyance guided using rope guides. A brief overview of these investigations is presented below.

The work (Bura 1970c) is a polemic of a thesis, popular in the 1960s and 1970s, originating in Great Britain and cited previously in this work, that arrangement of guide ropes on the long side of the conveyance is more favourable than corner arrangement. The reason of this situation was supposed to be a smoother run of conveyances, reflected in a smaller amount of shocks and lateral moves. The analysis of forces unbalancing the conveyance and setting it into motion seem to disprove British theory of side guide rope arrangement superiority. According to Bura (1970c), corner arrangement of guide ropes provides better stability of conveyance's ride, because ropes located in corners of the conveyance better resist the rotation of the conveyance. However, side arrangement might be considered better due to constructional issues (Bura 1970c).

Bura (1972) presents a thorough analysis of the airflow influence on the rope-guided conveyance, especially turbulent high-speed airflow in upcast mine shafts. Aerodynamic processes might impact on moving conveyance locally in different shaft sections, especially in the area of conveyances' passing and in the vicinity of the levels or ventilating channel, as well as continuously on the whole depth of the shaft, which is primarily caused by irregular distribution of air velocity in the shaft cross section, especially during movement of the conveyance and also by geometrical parameters of the conveyance. Effects of the described phenomena are lateral moves of the conveyance (Bura 1972).

Maximum deflections are caused by disturbances of the airflow due to passing of the conveyances and airflow in the vicinity of levels and ventilation channel. Calculations of aerodynamic forces acting on the cage allowed to compute maximum deflection of conveyances guided by ropes. Values of deflections obtained in calculations are shown in Table 1.2. It should be noted that there is no empirical verification of data presented in the table; it is only a result of analytical calculations.

Another issue of rope guidance and conveyance guided by ropes is its behaviour under the influence of additional factors, such as centre of gravity shift (Bura 1975b) or shaft lining displacements affecting the rope guidance geometry, caused by rock mass movement, due to mining operations carried out in shaft pillar (Bura 1975a). In both cases, the values of conveyance's lateral deflection might be increased. However, due to the theoretical approach to the issue, presentation of accurate calculations of these values is impossible. Especially, the case of shaft pillar liquidation requires individual approach, because of numerous factors affecting the shaft lining and rope guidance geometry.

An innovative, as for the 1970s, approach on empirical measurements is presented in a study by Bura (1971). GIG Research Institute proposed a method for rope-guided conveyance lateral movement measurements using video cameras. Unfortunately, no results of these investigations were published, so the practical aspects of this method cannot be verified. It was only stated that tests conducted in mine shafts were completed successfully.

A broad analysis of the behaviour of rope guides and rope-guided conveyance perpendicular to the shaft axis was described in a study by Slonina and Steuhler, 1980, which extensively presents an issue of rope guidance in European mine shafts

Table 1.2 The example of the aerodynamic force impact on deflection of the conveyance guided using elastic guidance (Bura 1972)

Phenomenon	*Maximum deflection with air speed equal, cm*			
	9 m/s		*12 m/s*	
	Cage	*Skip*	*Cage*	*Skip*
Ventilation channel inlet	2.2	2.1	4.0	3.8
Ventilation level inlet	4.0	3.4	7.0	6.0
Conveyances' passing	0.7	4.7	1.0	6.0
Kármán vortex street	0.25	0.85	0.30	1.00

operating in the 1970s. The authors analysed thoroughly lateral movements of the cage and stated that the behaviour of the cage in the direction of conveyance's ride is similar to the case of conveyance guided using stiff guidance.

A basis for lateral movement analysis is a characteristic of potential movement reasons, which are as follows (Slonina and Steuhler 1980):

- twisting moment of the hoisting rope,
- moments resulting from the eccentric position of the centre of gravity of the conveyance in relation to the axis of hoisting rope or the attachment point of the balance rope,
- forces and moments produced by guide ropes' deviation,
- forces and moments produced by hoisting ropes' deviation,
- Coriolis force,
- friction between the slipper and the guide rope,
- aerodynamic forces.

The main source of excitation of transverse and torsional movement of the conveyance is torsional (twisting) moment of the hoisting rope, according to Slonina and Steuhler, 1980. It acts directly on the conveyance. Reduction of its influence is possible in multi-rope winding installations, by using ropes with different directions of lay in suitable arrangement. In case of forces and moments produced by hoisting or guide ropes' deviation or eccentric centre of gravity position, it should be noted that conveyance's deflection might increase during its acceleration, deceleration, loading and unloading. It is an effect of a relationship between horizontal and vertical conveyance's oscillation, and processes listed above might be a source of vertical oscillations (Slonina and Steuhler 1980).

Aerodynamic phenomena are, besides the twisting moment of the hoisting rope, recognized the main reason of lateral oscillations of the conveyance. In addition to conveyance's movement, they are a reason of oscillation of ropes installed in the shaft. Uneven distribution of both lateral and vertical pressures on the conveyance gives rise to both quasi static and dynamic effects. A relatively large surface area of the conveyance results in oscillation rise in case of differential air velocities. The Coriolis results from the fact that the conveyance is moving perpendicular to the surface of the earth but not in the earth's axis of rotation. This is a quasistatic effect, whose magnitude in practice depends solely on the speed of the conveyance.

An accurate description of conveyance's behaviour guided by the ropes in the mine shaft is not an easy task. Conveyance's movement might be very complex, and it might comprise lateral moves, deflections or torsional motions. It should be noted that guide and hoisting ropes might be moved by forces acting on them, such as aerodynamic forces or forces caused by the conveyance's movement. It cannot be stated clearly which forces or moments are to be the key factors in particular cases of mine shafts, which additionally complicates forecasting of phenomena occurring in them (Slonina and Steuhler 1980).

Solutions and methods of rope-guided conveyance's deflection value calculation presented in the literature provide numerous approaches to the issue and emphasizes different aspects of the problem. One of them is a hypothesis (which is not verified tough) that the quantity of lateral movements decreases as the conveyance's velocity rises. Another states that conveyance's oscillations are in relationship with rope's parameter and its ability to vibration damping. There are also known differential equations describing conveyance's movement, presented by a Russian researcher. However, the principles of the described solution are not explained properly. Over the years, no complex theory fully describing the issue was presented. However, existing works allowed safe operation of the rope guidance in mine shafts (Slonina and Steuhler 1980).

Noteworthy is a conceptual work carried out in the R SA, where the characteristic of the domestic mining industry (extremely deep mines and extremely long hauling distances as a consequence) favours rope guidance in mine shafts. The issue of rope guidance was thoroughly analysed in the process of design of new mine shafts for the Palabora mine in the 1990s. For the purpose of development of open pit mine, which approached the end of its life, two shafts were sunk – 1,290 m deep production shaft with a diameter of 7.4 m and service shaft with a depth of 1,272 m and a diameter of 9.9 m. Skips operating in the production shaft, Mary Anne cage of the service shaft and counterweights of both cages are guided using guide ropes. The main cage is guided using stiff guidance. Due to lack of law regulations of technological gaps in mine shafts equipped with rope-guided conveyances in RSA at that time and known historic cases of conveyances' collision in different underground mines, it was decided to comprehensively analyse the case of rope guidance in designed shafts of Palabora mine, based on conducted tests and simulations (Taljaard and Stephenson 2000).

A foundation of the research was tests conducted in wind tunnel, which allowed to determine a value of factor, and this was used in calculations of side forces acting on the conveyance in motion. The next step was to determine the value of aerodynamic force caused by skips passing in the shaft. The listed parameters allowed to draw up a model, which was used for analysis of the maximum possible lateral deflection of conveyances, using different parameters of guide ropes. The value of deflection might be reflected in the designed area of the shaft cross section (Taljaard and Stephenson 2000).

The values of lateral deflection of skips obtained in tests described above are presented in Table 1.3 (Taljaard and Stephenson 2000).

Data presented in Table 1.3 show that rope torque has the biggest impact on conveyance's deflection. Other important factors are Coriolis force and aerodynamic force. The influence of buffeting caused by passing of conveyances is negligible (Taljaard and Stephenson 2000). Such results seem to confirm a thesis presented 20 years earlier (Slonina and Steuhler 1980), which state that it is impossible to unambiguously determine which force is the main source of conveyance's lateral deflection. However,

Table 1.3 Deflection of skips for various forces (Taljaard and Stephenson 2000)

Force	*Maximum Lateral Deflection of Skips*	
	Ascending Loaded (mm)	*Descending Empty (mm)*
Coriolis	63	53
Steady aerodynamic	30/26/19	25/19
Rope torque	90	82
Buffeting	2	4

according to Slonina and Steuhler (1980), the main source of conveyance's movement in the direction perpendicular to the shaft axis is the twisting moment (rope torque). Identical results were obtained in case of analysis of rope guidance in shafts of Palabora underground mine (Taljaard and Stephenson 2000).

To fully understand and describe the behaviour of the issue of rope guidance of conveyances in mine shafts, it is reasonable to investigate the phenomena occurring in guide ropes, which are the key elements of the rope guidance. Research on such phenomena was conducted in RSA and Australia and presented in the study by Greenway (2008). Sources of conveyances' lateral deflection described by this work are similar to those present in previous works (Slonina and Steuhler 1980; Taljaard and Stephenson 2000), which are as follows: aerodynamic forces, hoisting rope twisting moment and Coriolis force. Despite the small values of listed forces, they can significantly affect the behaviour of conveyance, because of the high elasticity of rope guides. Restoring forces, responsible for the return of the conveyance to its designed position, are significantly lower than in case of stiff guides. The author presents a thesis that it is necessary to properly investigate and understand the behaviour of laterally loaded guide rope, its shape and stiffness to fully understand and describe the behaviour of the whole rope guidance system (Greenway 2008).

Analysis of point load applied to a rope, resulting in its deflection, allows to determine the shape of the deflected rope. This shape is dependent on the point of the force application and the value of a "tension ratio" parameter, which is calculated based on tension in top and bottom ends of the rope and the weight of the guide rope. In case of low values of this parameter (low value of lower end tension), the deflected shape of the rope is catenary like, whereas for high values of tension ratio (high lower end tension), the shape is characteristic of taut string (Greenway 2008).

When considering the deflection of the rope in point of the force application, rope stiffness can be analysed. The shape of rope stiffness distribution is a function of the tension ratio mentioned before, while absolute stiffness depends on the rope's weight per metre. Stiffness at the top and bottom end of the rope approaches very large values but is only slowly varying in the mid-region of the rope (Greenway 2008).

Presented research on the behaviour of ropes (guide, hoisting and balance ropes) under the influence of forces acting laterally to ropes' axles might be helpful to understand and describe many phenomena occurring in mine shafts. Further research and analysis are needed; however, a solid base for them has been already built (Greenway 2008).

Particularly interesting are the results of investigation carried out in China, presented in works by Renyuan et al. (2014) and Renyuan, Zhu and Cao (2015, which provide complex and precise insight into the behaviour of rope-guided conveyances in mine shafts.

Renyuan et al. (2014) provide an analysis of conveyance's movement perpendicular to the shaft axis, made using the fluid-structure interaction method. Tests were conducted in vertical mining shaft of Yaoqiao mine, equipped with two rope-guided skips. Shaft diameter is 5 m, maximum velocity of skips is 9.7 m/s, weight of each skip is 15 tonnes and their loading capacity is 9 tonnes. Hauling distance in the shaft is only 320 m. Four guide ropes with a diameter of 40.5 mm are used for each skip. They are arranged in the corners of skips and tensioned using cheese weights. Technological gaps between conveyances are equal, i.e., 560 mm and between the conveyance and shaft lining it is 550 mm (Renyuan et al. 2014).

Simulations were carried out on 2D models, because it was considered accurate enough for the purpose of the research and 3D simulation computing demands are extremely high. For purpose of simulations, it was assumed that the centre of gravity of the conveyance is exactly below the point at which the hoisting ropes are attached to the conveyance. Rotation of the conveyance about a vertical axis is neglected in 2D simulation (Renyuan et al. 2014).

In simulations, conveyances are treated as rigid bodies moving along elastic guides. Lateral loads acting on conveyances vary with conveyances' vertical position and their longitudinal velocity. Similar to an earlier research, sources of conveyances, lateral movements are Coriolis and buffeting (aerodynamic) forces. However, Coriolis force was neglected in simulations, because in the analysed case, this force acts on conveyances in the eastern or western direction (depending on movement direction), and 2D analysis was carried out in the N-S direction (Renyuan et al. 2014).

Equations governing the behaviour of the rope-guided conveyance were solved by the ANSYS FLUENT program with PISO scheme, and user-defined function programmed by C language, with the modified Euler method. As the Coriolis force was neglected, forces acting on the conveyance are caused by aerodynamic phenomena. They reach extreme values while conveyances pass each other, so this case was thoroughly analysed. Acting of buffeting force, occurring during conveyances' passing is extended by turbulent airflow in the shaft (Renyuan et al. 2014).

Lateral aerodynamic buffeting force gives the conveyance a lateral impact acceleration, and then the conveyance begins to oscillate laterally. The full skip oscillates from about −34.2 to +38.5 mm while travelling upward, and the empty skip oscillates from about +45.1 to −65.5 mm while travelling downward. Moreover, the tendency of motion for conveyance is in agreement with the Chen's measurement on Yaoqiao vertical production shaft, conducted in 1979. However, values of oscillations measured by Chen were between 66 and 125 mm while travelling upward and between 1,354 and 150 mm while travelling downward (Chen 1979, 1985; Renyuan et al. 2014).

Renyuan, Zhu and Cao (2015) developed research presented above (Renyuan et al. 2014). FSI, the method used for simulations, is identical as in previously presented work. This time, uncompromisingly, a 3D model was used. Two mine shafts were analysed. First of them is equipped with two cages and another with a cage and counterweight. The parameters of these shafts are shown in Table 1.4.

Simulations conducted with ANSYS FLUENT software using user-defined function revealed that values of conveyances' lateral movements (both cages and counterweight towards each other) are significantly greater than their movements towards

Table 1.4 Parameters of hoisting systems in analysed mine shafts (Renyuan i in. 2015)

Parameter/Equipment	*Mine Elevator*	*Mine Cages*
Guide ropes	4 off for car, 2 off for counterweight, 24 mm	4 off per cage, 24 mm
Conveyance self-mass	Car – 2.1 Mg; counterweight – 2.6 Mg	1.6 Mg
Hoisting distance	62.4 m	62.4 m
Hoisting speed	4.0 m/s	4.0 m/s
Conveyance payload	1 Mg	0.8 Mg
Method of tensioning	Weight	Weight

Table 1.5 Values of side and lateral movements of conveyances (Renyuan i in. 2015)

	Car		*Counterweight*		*Car 1*		*Car 2*	
	Lateral	*Side*	*Lateral*	*Side*	*Lateral*	*Side*	*Lateral*	*Side*
Upper bound, mm	6.88	1.94	2.60	0.88	3.99	0.14	2.35	0.35
Lower bound, mm	3.13	2.63	2.15	0	1.03	0.63	2.37	0.04
Amplitude, mm	10.01	4.57	4.75	0.22	5.02	0.77	4.72	0.39

shaft lining. The values of oscillations are presented in Table 1.5 (Renyuan, Zhu and Cao 2015).

Summarizing the results of presented research on rope-guided conveyances, which was conducted during decades in different countries all over the world, it can be concluded that researchers agree on sources of conveyance's excitation. However, their views on the main reason of these deflections are different. It is worth noting that the most modern presented research conducted in the 2010s in China complies with results of the oldest presented tests, which were conducted by Bura in the 1970s in Poland. Both Renyuan's team and Bura state that aerodynamic forces are key sources of conveyance's oscillation. Other important sources of lateral movements of conveyance are Coriolis force (which was not analysed by Bura) and hoisting rope twisting moment, which were thoroughly analysed by researchers from RSA and Australia.

For decades of rope guidance presence in dozens of underground mines' shafts all over the world, many different technological solutions were constructed and applied. A brief overview of some of these solutions is presented below.

An interesting case, in terms of today's realia of mining industry and vertical transport, is a solution of rope guidance of conveyances applied in Consolidated Murchison mine in Northern Transvaal in RSA. One of the shafts operating in this mine was deepened to the depth of 426.72 m (distance of this extension was about 213 m) in the 1950s. Within the context of the issue of hoisting system with rope-guided conveyance operating on different mine levels described above, an interesting fact is that the shaft of the Consolidated Murchison mine operated on seven mine levels before deepening. After the shaft extension, another seven levels received connection with the shaft, so it could operate on 15 different levels (ground plus 14 mine levels) (Shippen 1958).

Before the deepening, mine shaft equipment comprised two cages guided with four guide ropes with a diameter of 22 mm each. Ropes were attached at the head frame and in the shaft sump below the seventh level. After the shaft extension, ropes were replaced with eight new guide ropes (four per cage) with a diameter of 30.2 mm and two rubbing ropes with a diameter of 31.8 mm. They were attached in the sump and on the head frame using coil springs, similar to the previous rope guidance construction (Shippen 1958).

Unfortunately, there is no information if the hoisting system could in fact operate on all of the mine levels, which are connected by the shaft. There is also no information about any equipment for conveyances' stabilizing applied in the shaft. However, it should be remembered that due to the rapid development of science and technology, 1950s' solutions are technical prehistory. It cannot be compared to modern solutions, because different mass transported in conveyances of mine shafts, archaic construction of winding machines and ropes and existing legislation, especially in the field of health and safety at work. Nevertheless, the example of the Consolidated Murchison mine's shaft can be a proof that rope guidance in shafts connecting many mine levels is not a previously unknown issue (Shippen 1958).

Stiff guidance of conveyances predominated hoisting systems in the European mining industry for decades. Great Britain was an exception. Rope guidance was in common use in the UK. In the late 1970s (1977), rope guidance was in use in 368 mine shafts in Great Britain and only in 20 in FRG (1979). Nowadays, due to vanishing of coal mines in Europe, analysis of types of guidance system in mine shafts is pointless. Therefore, a brief overview of historic data about rope guidance in European underground mines is presented below (Slonina and Steuhler 1980).

Slonina and Steuhler (1980) prepared an elaborate report comprehensively presenting an issue of rope guidance in European underground mines in the late 1970s. Within the framework of their research, the authors visited underground coal mines in Great Britain, France and West Germany. Data gathered in field visits in collieries point at the popularity of the weight tensioning of rope guides. Only in one of the presented mine shafts, hydraulic tensioning was used. In the majority of shafts, four half or full locked coil guide ropes with a diameter of at least 40 mm were used for the purpose of conveyance's guidance. Similarly, rubbing ropes were usually used. In five analyzed shafts, two were equipped with cages and three with skips. Winding distance in shafts was between 800 and 1,100 m (Slonina and Steuhler 1980).

Particularly interesting in terms of problem presented in this work are technological gaps and solutions of stiff guidance applied on mine levels and data about lateral moves of conveyances. In one of the shafts of French Merlebach coal mine, technological gaps were up to 600 mm, while in three different German mines, they were, respectively, 230–400 mm, about 350 mm and 220–460 mm. At the shaft stations, stiff guidance was applied, of construction potentially similar to solutions used in Polish collieries nowadays. The velocity of conveyances' entry on the stiff guidance was from 1 m/s (in Merlebach colliery), through 1.5 and 2.45 m/s (German mines) up to 3 m/s in Bevercots-Retford colliery in the UK and 4 m/s in Sigfries Giesen mine in FRG (Slonina and Steuhler 1980).

No accurate data on conveyances' lateral oscillations were gathered in shafts presented in the study by Slonina and Steuhler (1980). Only verbal descriptions of phenomena occurring in the shafts were available. The common elements of these descriptions are slight but noticeable oscillations during conveyances' passing. No precise observation was carried out though. In Merlebach mine, excessive lateral motions of conveyances were noted, but it was considered an effect of operation of another conveyance guided with stiff wooden guides beside rope-guided conveyances. Slight vibrations were also noticed during accelerating. In two German mines, lateral movements of conveyances caused by aerodynamic forces were noted. In one of them, it was an effect of airflow disturbances caused by shaft furniture and in another, it was an effect of ventilation channel influence (Slonina and Steuhler 1980).

German Siegfreid Giesen colliery reported a case of rope-guided conveyance collision in one of its foreshafts. However, technological gaps between the conveyance and foreshaft furniture in this case were only about 300–400 mm. Noteworthy is the single case of conveyance's rubs against the shaft lining in French Merlebach mine, because technological gaps in this shaft were up to 600 mm (Slonina and Steuhler 1980).

Obviously, the history of rope guidance of conveyances in European underground mines did not end in the 1970s. Despite closure of other coal mines in different countries, especially in the EU, numerous mines exploiting other materials operate with different solutions of rope guidance applied in their shafts. Such construction was used in a new shaft sunk in the 1990s for zinc and copper Pyhäsalmi mine, located in central Finland. This shaft was at that time one of the most modern and innovative mine shafts in the world. It also became the deepest mine shaft in Europe, and it still holds this title. In this 1,450-m-deep shaft with an inner diameter of 5m, a skip, its counterweight and a cage operate, all guided with ropes. Guide ropes are fixed suspended at the bottom of the shaft. The rope attachments in the head frame incorporate load cells to measure the rope tension. This can be adjusted with the help of hydraulic jacks. Simulations showed that all ropes should have the same pretension, contrary to common practice. Technological gaps in Timo shaft (that is the name of Europe's deepest shaft) are equal 400mm (ABB advertising materials).

Similarly, in RSA, rope guidance of conveyances is still in use. Based on simulations and tests presented above, all of the Palabora underground mine shafts, sunk in the 1990s, were equipped with a rope guidance system. According to the project, rope guides are applied in production shaft, where four rope-guided skips operate and in service shaft, where guide ropes were applied for main cage's counterweight and for both Mary Anne cage and its conveyance. Four rope guides were applied for each of skips, Mary Anne cage and main cage's counterweight. Mary Anne cage's counterweight is guided with only two ropes. Guide ropes are hydraulically tensioned, using a system assembled on the head frame (Taljaard and Stephenson 2000).

Limited availability of literature about rope guidance of conveyances in mine shafts in North America does not mean that such a type of guidance is not in use there. The situation is quite the opposite, which is indicated by observation of European researchers, general mining technology literature or numerous catalogues of the element of rope guidance systems, such as ropes of special type, suspensions or rope tensioning systems (de la Vergne 2008; advertising materials of NorthernStrands; advertising materials of FLSmidth).

Polish experiences in rope guidance of conveyances in underground mine shafts, especially in coal mines, were comprehensively described by Lesław Bura. According to his works, released in the early 1970s (Bura 1970b, 1970c), only in 5% of mine shafts in Polish collieries equipped with hoisting systems, rope guidance of conveyances was applied. In practice, it was about 20 shafts. However, there were shafts where only counterweight was guided by ropes and a cage or a skip was guided using stiff wooden or steel guides. The presented number of 20 shafts does not contain systems of rope guidance used in numerous shafts sunk at that time (Bura 1970b).

Among 20 shafts of coal mines, where rope guidance was used, only three of them were mining shafts equipped with skips. Other shafts were man, material or ventilation shafts. Over 60% of rope guidance solutions were applied in shafts with depth up to 300m and diameter up to 5,500mm. The deepest one was 800m deep and the shallowest only 125m. The diameter of none of these shafts exceeded 6,600mm. Every shaft of Polish colliery, where rope guidance was applied, was a one-compartment shaft, with little exceptions, equipped with ladder compartments. These exceptions were two shafts, of which one of them was equipped with auxiliary hoist and another was not equipped with neither ladder compartment nor auxiliary hoist (Bura 1970b).

The most popular arrangement of rope guides in mine shafts in Poland at the turn of 1960s and 1970s was a corner arrangement of four guide ropes. However, side arrangement was also used. Two-rope systems were also spotted, but they were used only for guiding counterweights and Mary Anne cages. The diameters of guide ropes were between 22 and 42 mm, and they were mostly full locked coil ropes. Single braid ropes were in the minority. Tensions of the ropes were between 8 and 25 kN/100 m of the rope in hoisting systems with multiple hoisting ropes and between 32 and 49 kN/100 m of the rope in hoisting systems with single hoisting ropes (Bura 1970b).

Data on rope tensioning might be found interesting. Less than half of Polish shafts equipped with rope guidance at that time were also equipped with cheese weights for ropes' weight tensioning. In 50% of shafts, systems of spring tensioning were applied. In one remaining shaft, ropes were weight tensioned by a system of cheese weights installed on the head frame with special leverage construction (Bura 1970b).

Analysis of different examples of rope guidance solutions in mine shafts of Polish collieries revealed a vast span of used technological gaps between conveyances or between conveyances and shaft equipment or its linings, especially extremely low values of these gaps in some cases. The smallest technological gap between conveyances was only 155 mm, while the biggest was 755 mm. Furthermore, the smallest technological gap between the conveyance and shaft furniture was 125 mm and between the conveyance and shaft lining only 100 mm. The greatest values of technological gaps were 580 mm between the conveyance and shaft furniture and 900 mm between the conveyance and shaft lining. The listed values are surprising in terms of technological gaps in other European mine shafts presented above. However, it should be noted that those European shafts were significantly deeper. Also, velocities of conveyances in these shafts and their loading capacities were much greater. Nevertheless, technological gaps in Polish shafts equipped with rope guidance at the turn of the 1960s and 1970s were smaller than their values suggested by the Ministry of Mining and Energetics, and in four cases, they failed to comply with the law existing at that time. It was obviously associated with problems with hoisting systems' operation. A solution of this issue was using high values of tension in ropes. In one case, it was necessary to assemble baffle plates at the whole depth of the shaft to prevent conveyance from colliding with elements of ladder way compartment (Bura 1970b).

Nowadays, rope guidance of conveyances is used in 16 shafts of Polish underground mines. Although this number is similar to the amount of rope guidance systems in 1970 (Bura 1970b), currently only three of these shafts operate in coal mines. The remaining 13 hoisting systems equipped with rope guidance are used in copper mines of KGHM Polska Miedź SA. So, reduction of rope guidance applications in Polish collieries is noticeable. However, a total number of mine shafts significantly reduced over the years as a consequence of closing down dozens of coal mines. The common use of rope guidance of conveyances in copper mines of Pre-Sudetic Monocline is caused by significant depth of shafts, long service life, high durability and reduction of airflow resistance, which is particularly important because of the requirement for huge amounts of fresh air in mine workings. Long experience of Lower Silesian engineers working for KGHM indicates validity of rope guidance application in copper mines' shafts because of easy and safe maintenance, quick replacement of guide ropes, no collisions of conveyances and no occurrence of any other issues connected with hoisting system operation, which allows to state that application of rope guidance is economically and practically reasonable (Tobys and Tytko 2011; Olszyna et al. 2018).

To sum up the overview of rope guidance solutions in Poland and other countries, there is no record of a hoisting system equipped with rope guidance of conveyances operating on more than one mine level in the literature. Existing law regulations involve occurrence of such situation and requires application of special shaft chairing constructions and stiff guidance in particular. Numerous problems associated with such constructions make such solutions unpopular.

However, there are historic records of hoisting systems equipped with rope guidance of conveyances operating in mine shafts connecting a number of mine levels. As shown above, Consolidated Murchison mine's shaft from RSA, operating in the 1950s (Shippen 1958) might be an example of such shaft. It cannot be treated as a valid example in terms of today's state of technology.

According to Slonina and Steuhler (1980), loading and unloading of conveyances might be conducted on mid-levels, while specific retractable holding devices are used on shaft stations to stabilize the conveyance. For reasons of safety, they have to be synchronized with systems of conveyance's movement control and monitoring of its position in the shaft. The authors provide only principles, but they do not present any particular solution. However, such specific requirements for a holding device presented by the authors might suggest that a similar system was in use (Slonina and Steuhler 1980).

In the 1970s, in the USA, a decking device for mine cages and the like conveyances was patented. As the author states, as an empty cage is loaded at a deep level, it gets progressively heavier and therefore descends in the shaft under the influence of the stretch. Similarly, as the cage is unloaded at this station, it gets progressively lighter, and it consequently rises in the shaft under the influence of the cable stretch. To prevent such situations, a holding device was designed. A conveyance is stabilized using a system of jaws moved by hydraulic unit. They engage a structure on the conveyance to immobilize it in the shaft (Trollope 1965).

It is not the only patented solution of stabilizing conveyances. A problem of rope stretching was noticeable, and so many devices were designed to solve it, such as in Kramer (1982) and Watt (1972). These devices are characterized by similar idea of work and possible application, but they differ in approach to the problem. There is no record of application of these devices and conditions of possible application. Thus, it is impossible to determine if they were ever used to stabilize a conveyance on the shaft station.

A device of similar idea of work was used in the Timo shaft of Pyhäsalmi mine. The shaft connects ground level with only one mine level, but it is equipped with service station for the purpose of maintenance of ropes and conveyances. The station is equipped with hydraulic lifting devices, which is used to hold conveyances to unload head ropes. The details of this solution were not presented (ABB advertising materials).

There are different hydraulic lifting devices available on the market, designed and produced by numerous producers of mining equipment. All of these devices are used for the purpose of skips' service during their loading and unloading (advertising materials of FLSmidth, Zitron, Voith). Despite the idea of utilization of hydraulic device for lifting and holding a conveyance is similar to previously presented device, construction of these devices is different. There is also no device for stabilizing conveyances at mid-levels available on the market.

REFERENCES

ABB Mine Hoisting System at Pyhasalmi Mine – advertising materials of ABB.

Adamczyk A. (2012) *Kopalnia Charlotte. Dzieje KWK Rydułtowy-Anna 1806–1945*; wydane przez ART. DRUK, Rydułtowy 2012 [In Polish].

B-1500/Ex/AC-2m/s hoisting machine. Innovative solution for use in underground excavations in the methane and coal dust explosion hazard zones as well as in safe zones – advertising materials of MWM Elektro sp. z o.o.

Bojarski P, Bulenda P, Kamiński P. and Nowak J. (2019) *Innowacyjny sposób stabilizacji naczyń wyciągowych wyciągu szybowego z prowadzeniem linowym na poziomie pośrednim*, in proceedings of International Mining Forum 11–12.04.2019, Katowice, Poland [In Polish].

Bura L. (1970a) *Drgania swobodne nietłumione naczynia wyciągowego na prowadnikach linowych*, Komunikat GIG nr 485, Wydawnictwo „Śląsk", Katowice 1970 [In Polish].

Bura L. (1970b) *Linowe prowadzenie naczyń wyciągowych w szybach krajowych*, Komunikat GIG nr 483, Wydawnictwo „Śląsk", Katowice 1970 [In Polish].

Bura L. (1970c) *Wpływ rozmieszczenia lin prowadnikowych na ruchy naczynia wyciągowego w szybie*, Komunikat GIG nr 486, Wydawnictwo „Śląsk", Katowice 1970 [In Polish].

Bura L. (1971) *Zastosowanie kamer filmowych do wyznaczania w szybie poziomego ruchu naczynia wyciągowego prowadzonego na linach*, Komunikat GIG nr 517, Wydawnictwo „Śląsk", Katowice 1971 [In Polish].

Bura L. (1972) *Warunki pracy i projektowanie linowego prowadzenia naczyń wyciągowych w szybach wydechowych*, Komunikat GIG nr 565, Wydawnictwo „Śląsk", Katowice 1972 [In Polish].

Bura L. (1975a) *Prowadzenie linowe naczyń wyciągowych w szybach w czasie eksploatacji ich filarów ochronnych*, Komunikat GIG nr 657, Wydawnictwo „Śląsk", Katowice 1975 [In Polish].

Bura L. (1975b) *Wpływ poziomego przesunięcia środka ciężkości naczynia wyciągowego na warunki pracy jego prowadzenia linowego*, Komunikat GIG nr 647, Wydawnictwo „Śląsk", Katowice 1975 [In Polish].

Carbogno A. and Żołnierz M. (2009) Zamocowanie na wieży wyciągowej prowadników linowych za pomocą zacisków klinowych, *Transport szybowy 2009* – Monografia, s. 151–167, ITG-KOAG, Gliwice 2009 [In Polish].

Chen X. (1979) Analysis of the Soviet calculation formula for the clearance between rope-guided conveyance. *Design of Coal Mine* 26(4), s. 17–23.

Chen X. (1985) Swing of hoisting conveyance using steel rope guides. *Coal Science and Technology*, s. 23–26.

Czaja P. and Kamiński P. (2017) *Wybrane zagadnienia techniki i technologii głębienia szybów.* Kraków: Szkoła Eksploatacji Podziemnej — (Biblioteka Szkoły Eksploatacji Podziemnej). — Bibliografa. przy rozdz. — ISBN: 978-83-927920-9-3.

Czaja P., Kamiński P., Olszewski J. and Bulenda P. (2018) *Polish experience in shaft deepening and mining shaft hoists extending on the example of the Leon shaft IV in the Rydułtowy mine //* World Mining Congress [online document]: [19–22 June], Astana, Kazakhstan.

de la Vergne J.N. (2008) *Hard Rock Miner's Handbook*, Stantec Consulting Ltd., Edmonton.

FLSmidth: Mine shaft systems – advertising materials of FLSmidth.

Greenway M.E. (2008) Lateral stiffness and deflection of vertical ropes with application to mine shaft hoisting. *Australian Journal of Mechanical Engineering*, 5(1), s. 59–70.

Jaros J. (1984) *Słownik historyczny kopalń węgla na ziemiach polskich*, Śląski Instytut Naukowy, Katowice 1984 [In Polish].

Kamiński P. (2018) *Zmiana funkcji wyciągu pomocniczego na wyciąg mały w szybie Leon IV KWK ROW ruch Rydułtowy*, dissertation, Department of Cable Transport WIMiR, AGH UST, Kraków 2018 [In Polish; unpublished].

Kamiński P. (2020) *Polish Experience in Shaft Deepening and Mining Shaft Hoists Elongation*, Mining Techniques-Past, Present and Future, InTech.

Kostrz J., Olszewski J., Czaja P., Deja J. and Witosiński J. (2000) Zastosowanie betonów odpornych na silną agresję siarczanową i magnezową w budownictwie podziemnym, *Budownictwo Górnicze i Tunelowe* 3/2000 [In Polish].

Kowal L. (2013) Nowoczesne maszyny wyciągowe i ich wyposażenie na przykładzie efektów współpracy ITG KOMAG z przemysłem, *Maszyny górnicze* 2/2013, p. 61–70 [In Polish].

Kramer G. (1982) *Means for holding means of transport associated with an extraction well*; FR 2506281A1.

Madej M., Radowski, R., Kowal L., Turewicz K. and Helmrich P. (2013) Nowa maszyna wyciągowa B-1500/EX/AC-2m/s przeznaczona do stosowania w przestrzeniach zagrożonych wybuchem, *Transport szybowy 2013*- Monografia, s. 201–214, ITG-KOMAG, Gliwice 2013 [In Polish].

Mańka E., Słomion M. and Matuszewski M. (2018) Constructional features of ropes in functional units of mining shaft hoist. *Acta Mechanica et Automatica* 12, s. 66–71

Northern Strands: mining wire rope attachment and equipment – advertising materials of Northern Strands

Nowak J, Bulenda P, Piszczan Z and Kamiński P (2019) *Innowacyjny sposób stabilizacji klatki wielkogabarytowej wyciągu podstawowego szybu Leon IV*, in proceedings of Mechanizacja, automatyzacja i robotyzacja w górnictwie. VI międzynarodowa konferencja, 12–14.06.2019, Wisła, Poland [In Polish].

Olszewski J., Bulenda P., Chomański B., Jara Ł. and Kamiński P. (2017) Szyb Leon IV – pogłębianie i wydłużanie górniczych wyciągów szybowych, *Inżynieria Górnicza: kwartalnik specjalistyczny*; ISSN 2353-5490. – 2017nr 1–2, s. 49–52.

Olszewski J., Czaja P., Bulenda P. and Kamiński P. (2018) Pogłębianie oraz wydłużanie górniczych wyciągów szybowych szybu Leon IV w kopalni KWK ROW ruch Rydułtowy, Przegląd górniczy vol. 74, nr 8, s. 7–16.

Olszyna G., Tytko A. and Tobys J. (2018) Eksploatacja lin prowadniczych i odbojowych, Napędy i Sterowanie; ISSN 1507–7764. – 2018R. 20 nr 4, s. 110–114. [In Polish].

Renyuan W., Zhu Z. and Cao G. (2015) Computational fluid dynamics modelling of rope-guided conveyances in two typical kinds of shaft layouts, *PLoS One*. 10, e0118268.

Renyuan W., Zhu Z., Chen G., Cao G. and Li W. (2014) Simulation of the lateral oscillation of rope-guided conveyance based on fluid structure interaction. *Journal of Vibroengineering*, 16, (3), s. 1555–1563.

Ryndak P. and Kowal L. (2015) Efekty współpracy ITG KOMAG z firmą MWM Elektro sp. z o.o., *Maszyny górnicze* 3/2015, s. 53–61 [In Polish].

Shippen J.W. (1958) Installing guide ropes in a vertical shaft. *Journal of the Institution of Certified Mechanical and Electrical Engineers*, South Africa, January.

Slonina W. and Stuehler W. (1980) *Safety problems posed by rope shaft guides, research report*, Commission of the European Communities, Mines Safety and Health Commission, Luxembourg.

Taljaard J.J. and Stephenson J.D. (2000) State-of-art shaft system as applied to Palabora underground mining project. *The Journal of The South African Institute of Mining and Metallurgy*, 11–12/2000, s. 427–436.

Tobys J. and Tytko A. (2011) Eksploatacja lin prowadniczych i odbojowych w górniczych wyciągach szybowych, *Transport szybowy 2011* – Monografia, s. 375–383, ITG KOMAG, Gliwice 2011 [In Polish].

Trollope W.G.A. (1965) *Current limiting chairing system*; US 3477548A.

VOITH: Key components for shaft hoisting – advertising materials of VOITH.
Watt I. (1972) *Decking device for mine cages and the like conveyances*; US 3800918.
Wichur A., Frydrych K. and Kamiński P. (2015) *Static calculations of mine shaft linings in Poland (selected problems)* // in proceedings of Vertical and decline shaft sinking – good practices in technique and technology: International Mining Forum 2015: 23–27 February.
Wowra D., Nowak J., Witkowski J., Izydorczyk P. and Kamiński P. (2017) *Wydłużenie Górniczego Wyciągu Szybowego – Szyb Leon IV* // In: IV PKG [online document]: IV Polski Kongres Górniczy 2017: 20–22.11.2017, <atl>Kraków [In Polish].
Wowra D., Nowak J., Witkowski J., Ratuszny K., Chomański B. and Kamiński P. (2018) Nowoczesne rozwiązania infrastruktury okołoszybowej – urządzenia pomocnicze na poz. 1150m szybu Leon IV, in proceedings of Szkoła Eksploatacji Podziemnej, 26–28.02.2018, Kraków, Poland [In Polish].
Wysocka-Siembiga A. (2016a) Polska Grupa Górnicza stała się rzeczywistością, *Gazeta firmowa PGG* 01/2016, s. 3–6 [In Polish].
Wysocka-Siembiga A. (2016b) Szyb Leon IV gotowy, *Gazeta firmowa PGG* 4–5/2016, s. 10–11 [In Polish].
Zitron: mining shaft equipment – advertising materials of Zitron.

Chapter 2

Description of the retractable guidance system

2.1 CHARACTERISTICS OF THE RETRACTABLE GUIDANCE SYSTEM

The retractable guidance system replaced the chairing mechanism of a rope-guided cage at level 960 of Leon IV shaft in the Rydułtowy Coal Mine. This system was originally introduced in the project Projekt techniczny (2018), and it was developed to solve the problem of the cage guidance through the level 960. The main reason for the development of this system was to reconcile safety with effectiveness. To provide a proper level of safety using the chairing system, it was necessary to reduce the speed of the cage from 10 to 0.5 m/s, 100 m before the level 960. The effect of such situation was significant extension of a single ride time to or from the level 1,150 m.

Stiff guides on the level 960 m were replaced with the retractable guidance system, which consist of two pairs of top and two pairs of bottom guides and a supporting structure, comprising four supporting frames assembled to the shaft lining by additional beams, structurally independent of the existing shaft chairing (Wowra et al. 2017; Olszewski et al. 2018; Kamiński 2020).

Retractable guides can be moved between two positions: resting and working position. The motion is restricted by:

- an element of the supporting frame, named the race – bottom section;
- two articulated links (top and bottom) – top section.

Each of the bottom sections of the retractable guidance system are driven by a hydraulic cylinder, assembled to the guide's section and the supporting frame. Top sections of the retractable guidance system are driven by two hydraulic cylinders each.

During the movement of the conveyance between the levels 1,150 and zero, all sections of the retractable guidance system are set in the resting position (Figure 2.1).

The retractable guidance system set in the resting position allows the conveyance to move through the level 960 with its full speed, which is equal 10 m/s. The system is also set in the resting mode when the cage is moved between levels 960 and 1,150 in both directions (Kamiński, Prostański and Dyczko 2021).

The system is set in the working mode when the cage stops at the level 960. It is done by a banksman on the level 960. Sections of the retractable guidance system in the working position stabilize the conveyance on the level. Sections of the system are inserted between the elements of the guiding shoes of the cage. Sections of the

DOI: 10.1201/b22695-2

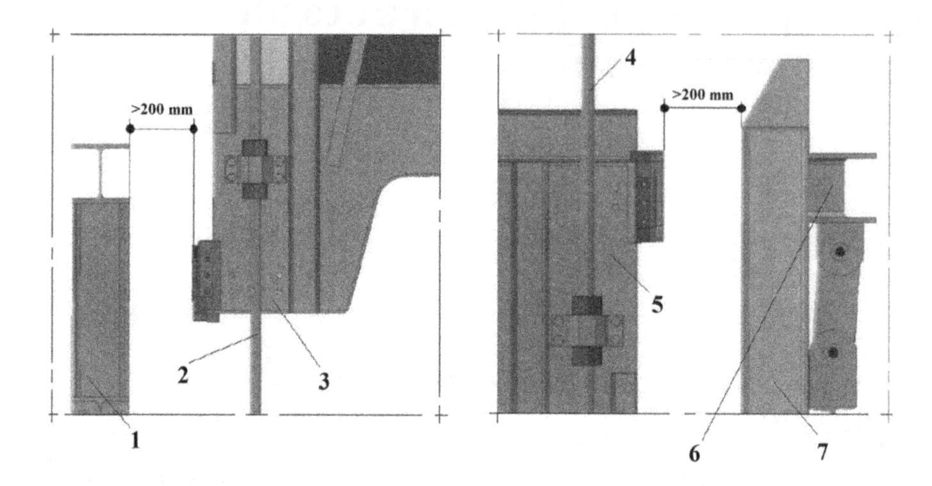

Figure 2.1 Sections of the retractable guidance system in resting position; 1 – supporting frame of the bottom section, 2, 4 – rope guides, 3 – bottom transom of the cage, 5 – top transom, 6 – supporting frame of the top section, 7 – top section of the retractable guidance system.

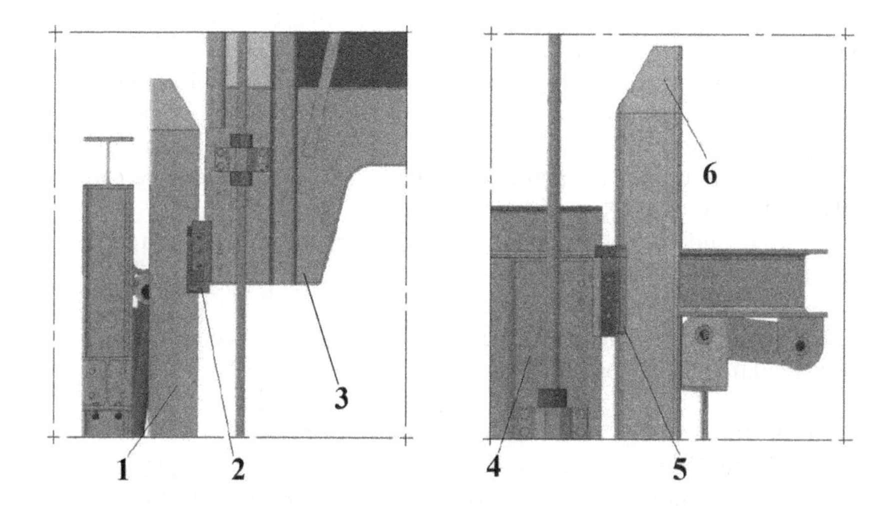

Figure 2.2 Sections of the retractable guidance system in working position; 1 – bottom section of the retractable guidance system, 2 – guiding shoe, 3 – bottom transom, 4 – top transom, 5 – guiding shoe, 6 – top section of the retractable guidance system.

retractable guidance system in working position are presented in Figure 2.2 (Czaja et al. 2018; Projekt techniczny 2018; Kamiński, Prostański and Dyczko 2021).

Layout of the retractable guidance system is shown in Figure 2.3. The sections of the system comprise 180 × 260 mm box beams made of C260 profiles. Supporting frames, made of HEB 260, are attached to two technological beams with M24 Hex bolts, class 8.8.

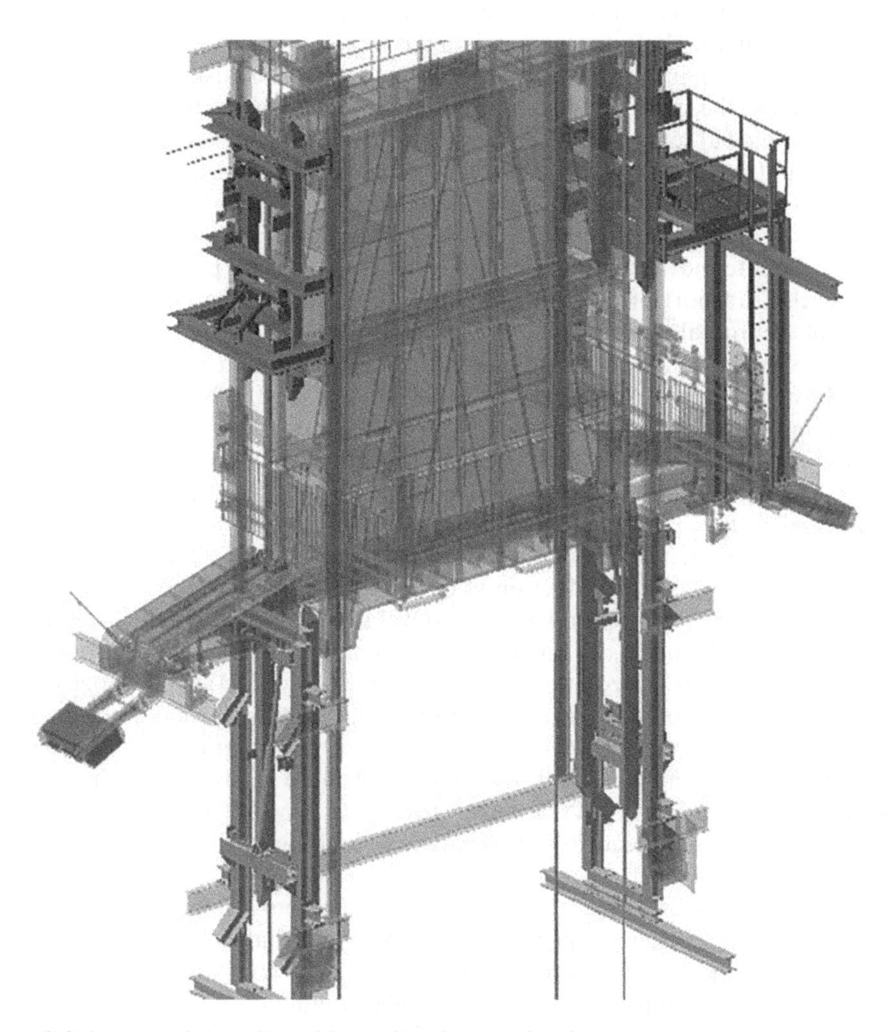

Figure 2.3 Layout of the adjustable guiding (grey colour).

Rope guidance is a common way of guiding a conveyance in the mine shaft. A conveyance is moved using one or multiple hoist ropes and guided using guiding ropes. They are attached to the head frame and stabilized in their bottom ends in the shaft sump. Elastic guidance of conveyances is most commonly used in the mine shafts where the conveyance is moved between ground level and only one mine level.

A decking device for mine cages and conveyances (Watt 1972) is also used in various mine shafts. Its task is to block a cage in the position and stabilize it, especially on mine levels. This device has two jaws movable in a guide towards each other to a first position in which they engage a structure on the conveyance to immobilize the same in the shaft, and away from each other through a second position in which one of the jaws is in contact with the structure to a third position in which the jaws are retracted and can no longer contact the structure. The jaws are driven between the first and third positions at a controlled speed to eliminate any tendency for the conveyance to move uncontrollably in the shaft due to stretch in the cable.

There is a Polish invention, called the frontal guide (Czekalski and Kubicki 1988), which is used at the inter-levels of mines in mine shafts where stiff guidance is used. The frontal guide at inter-levels is an element of the stiff guidance of the conveyance. On the level, it is permanently attached to a frame, which is rotary hinged to a construction of the shaft station. It is also detachably jointed with a locking device to stiff guides assembled in the shaft at both ends. The construction of the frontal guide allows the cage to move through the inter-level. It also allows people to get past of it on the level. Transport of people or materials, especially using mine carts, requires opening of the frontal guide.

Also a solution of the rope guidance in the mine shaft with at least two stations is known. In such situation, stiff guidance on the levels is used. It is necessary to provide proper stabilization of the conveyance on the shaft station. Such construction, used on shaft stations, usually consists of four guides, one for each of the conveyance's corner. Before entering the construction of stiff guidance at the level, the conveyance's speed has to be reduced.

The goal of constructors of the retractable guidance system was to solve the problem of necessity of the reduction of conveyance's speed in the vicinity of shaft stations in the shaft with at least one inter-level in which conveyances is guided using rope guides. The necessity for speed reduction is caused by the way of entering the stiff guidance assembled at the shaft station. Such reduction of velocity causes time extension of people or materials transport. Moreover, rope-guided conveyance tends to move laterally. Such lateral moves are also magnified by conveyance's breaking and turbulent airflow in the vicinity of mine levels. All of these factors make the process of entry of the conveyance onto the stiff guides very hard. In extreme cases, they can lead to a conveyance's blocking in the construction of the stiff guidance.

Presented invention of the retractable guidance system is intended to use in mine shafts with hoisting systems, operating on at least two shaft stations (and a ground level) in which conveyance is guided using rope guides. The conveyance in this system is equipped with typical guide shoes for elastic guidance. On the shaft station, supporting frame is assembled. It is equipped with the top and bottom section of the retractable guidance system on each side of the conveyance.

Top and bottom transom of the conveyance are equipped with typical guide shoes. There is no need for the assembly of additional equipment.

Sections of the retractable guidance system work as frontal guides. They are connected with supporting frames using hydraulic cylinders. Swinging bridges are assembled on the shaft station.

The key feature of the retractable guidance system is a possibility of moving the conveyance through an inter-level with its full velocity, without the need for any speed reduction in the vicinity of the shaft station. In case of moving the conveyance through an inter-level retractable guides are in the resting (idle) position, they are moved towards the shaft lining to provide sufficient gap between their construction and the conveyance. This gap, according to the Polish law, is equal to 200 mm. In such situation, there is no need for entering the conveyance onto the stiff guides; thus, lateral moves of the conveyance are limited, and there is no risk of conveyance's blocking.

When people or materials are transported to the level with the retractable guidance system assembled, its top and bottom sections are moved towards the conveyance after it stops on the shaft station. The retractable guidance system cooperates with guide shoes of the conveyance. It allows to perform transport operations to this level with no limits, also using swinging bridges assembled on the level.

The presented system allows the conveyance to move with its full speed throughout the length of the shaft, until the final station. Full speed of the conveyance is also possible in the vicinity of shaft stations, which is not possible in case of using typical stiff guidance at the levels.

Construction of the retractable guidance system is presented in the following figures: Figure 2.4 shows the diagram of the rope guidance in the mine shaft, Figure 2.5 presents the retractable guidance system at the inter-level in the idle position and Figure 2.6 – the retractable guidance system in the working position (Bulenda and Kamiński 2020).

Rope guidance of a conveyance is installed in a mine shaft with two inter-levels, i.e., two shaft stations and one bottom level. A conveyance is guided using rope guides. A hoist rope is attached to the conveyance. It is set in motion by a hoist installed on the surface. Guide ropes are assembled in the head frame with suspensions. Their bottom ends are stabilized with cheese weights. Rope guides cooperate with guide shoes, assembled on top and bottom transoms of the conveyance. Their task is to stabilize the conveyance in plane perpendicular to shaft axis.

A bottom level is equipped in the shaft chair and stiff guides, which is not presented in details together with equipment required for handling transport of people and materials. Various constructions of the shaft chair are applied.

Each of inter-levels is equipped with a retractable guidance system, comprising supporting frames and sections of retractable guides. Supporting frames are joint to the construction of the shaft chair. Sections of the retractable guidance system are connected with the supporting frames with hydraulic cylinders. Those cylinders are also responsible for changing the position of the sections from idle to working mode. Inter-levels are also equipped with swinging bridges, which are used for communication between the level and the conveyance.

Inter-levels are also equipped with typical elements of the shaft station, which are safety platforms used for maintenance and repairs of guidance system elements. Additionally, a control room is installed on every shaft station equipped with the retractable guidance system. Devices used for control of hydraulic cylinders are placed in this room.

The construction of the retractable guidance system allows setting them into resting position, retracted towards the shaft lining. In such situation, conveyance is able to move through the inter-level with its full speed, e.g., between ground and bottom level, when conveyance breaks before the bottom level. The retractable guidance system in the resting position provides sufficient technological gap between the shaft elements, which is required by Polish law (Bulenda and Kamiński 2020).

Construction of the retractable guidance system allows also setting them into working position. In this case, retractable guides cooperate with guide shoes of the conveyance. If transport of people or materials is carried out to the inter-level, immediately after conveyance stops, retractable guides are set into working position by moving them towards conveyance's guide shoes. Retractable guides in working position stabilize the conveyance on the shaft station.

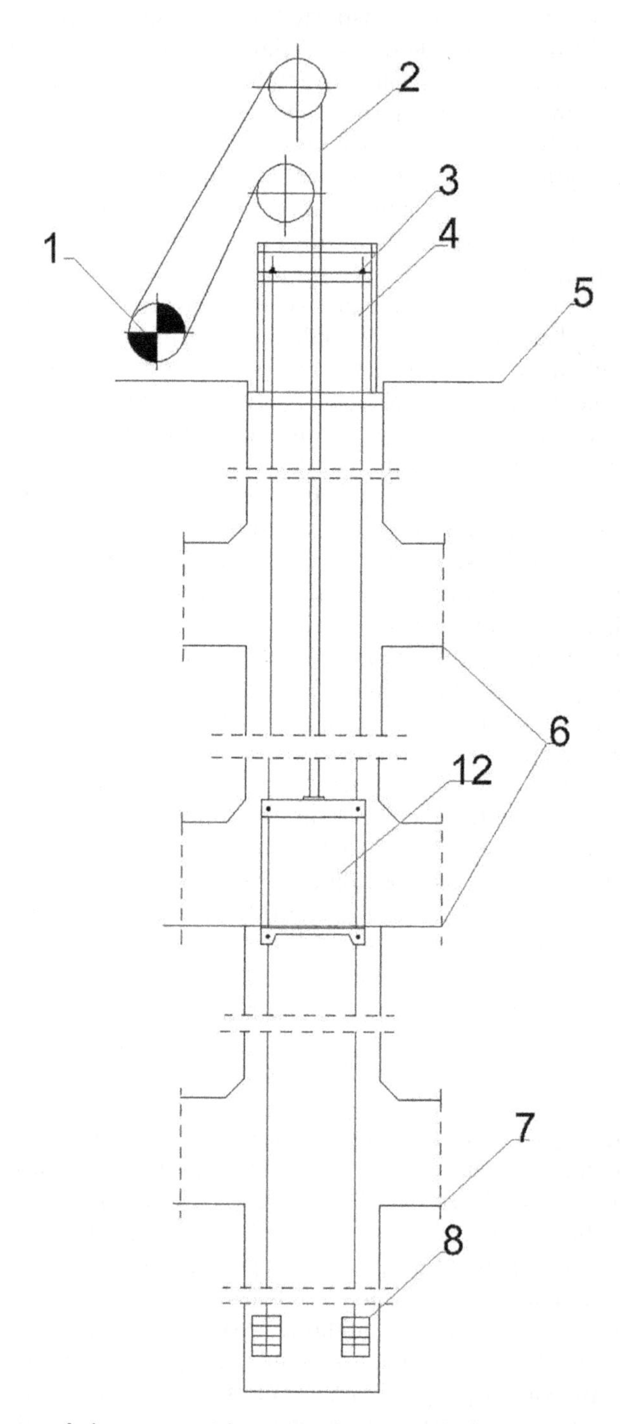

Figure 2.4 Diagram of the rope guidance in the mine shaft; 1 – hoist, 2 – hoist rope, 3 – guide rope suspensions, 4- guide ropes, 5 – ground level, 6 – inter-levels, 7 – bottom level, 8 – cheese weights, 12 – conveyance.

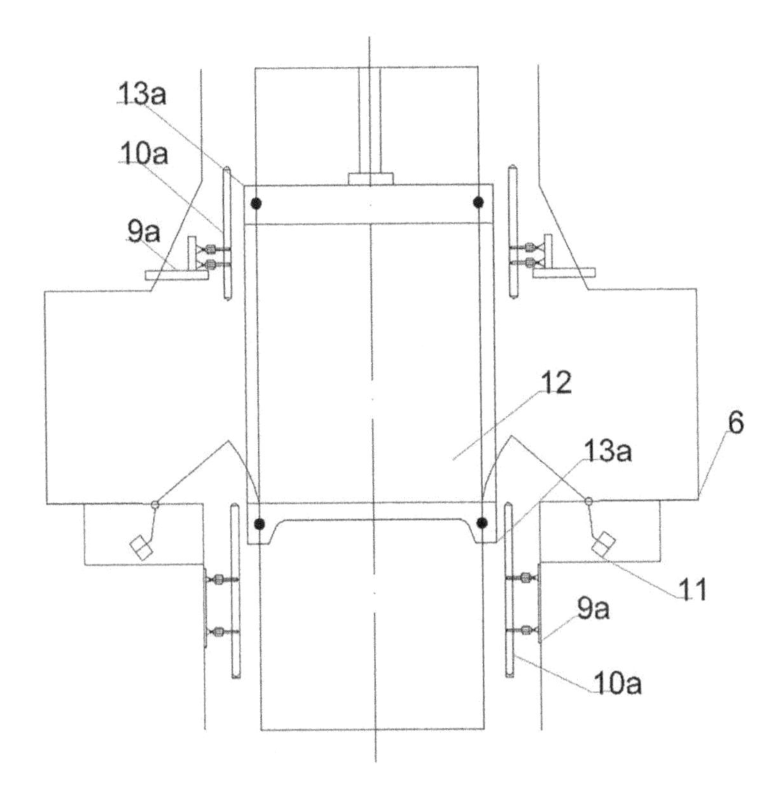

Figure 2.5 Diagram of the shaft station equipped with the retractable guidance system – retractable guides in resting position; 6 – inter-level, 9a – top supporting frame, 9b – bottom supporting frame, 10a – top section of the retractable guidance system, 10b – bottom section of the retractable guidance system. 11 – swinging bridges, 12 – conveyance, 13a, 13b – guide shoes.

One of the most common types of conveyances used in underground mines is a multi-deck cage. In case of using this type of the cage and necessity for moving the cage to align its different decks to the level of the level's floor, maneuvering is conducted with the retractable guides in working position, moved towards the guide shoes of the cage (Bulenda and Kamiński 2020).

Static analysis of the construction is presented in the forthcoming subchapter. Also, measurements of forces acting on the construction of the cage as well as stress in cage's bridle hangers were conducted. Results of these tests are presented in the next chapter.

Power supply and control system of the retractable guidance system is a unit of electric and hydraulic devices, the task of which is to transfer hydraulic medium to actuators responsible for movement of the guides and control their state and position. The system consists of two main parts located symmetrically on the eastern and western side of the shaft stage: bottom section comprising two elements stabilizing bottom transom of the cage and top section stabilizing top transom of the cage. For the purpose of safety of hoisting systems operating in Leon IV shaft, the system communicates with their signaling systems.

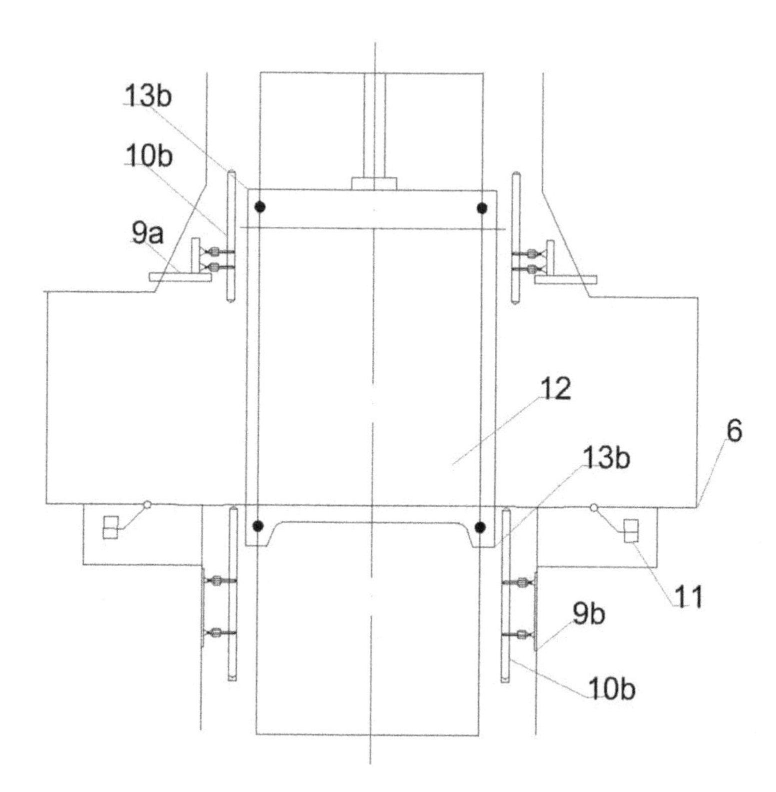

Figure 2.6 Diagram of the shaft station equipped with the retractable guidance system – retractable guides in working position; 6 – inter-level, 9a – top supporting frame, 9b – bottom supporting frame, 10a – top section of the retractable guidance system, 10b – bottom section of the retractable guidance system. 11 – swinging bridges, 12 – conveyance, 13a, 13b – guide shoes.

Power supply and control system comprises the following (Bojarski et al. 2019):

- control unit [S] – electric device, which generates necessary control signals based on the retractable guides' position, work of hydraulic aggregates and control panel.
- Control panel [ST] – panel assembled at the workstation of signaler used to control hydraulic aggregates and retractable guides.
- Auxiliary panel [SP] – panel accessible from the conveyance on the level 1,000, used to control retractable guides by signalman in case of using shaft revision or self-ride mode.
- Hydraulic aggregates [Ah] – devices equipped with hydraulic elements necessary to generate hydraulic medium pressure, its regulation and distribution to actuators.
- Retractable guides [P] – mechanical devices moved with actuators. Their position is controlled using sensors, the state of which determines the work of the control system and shaft signalling system.
- Shaft signalling system [Sz] – receives signals generated by sensors of retractable guides.

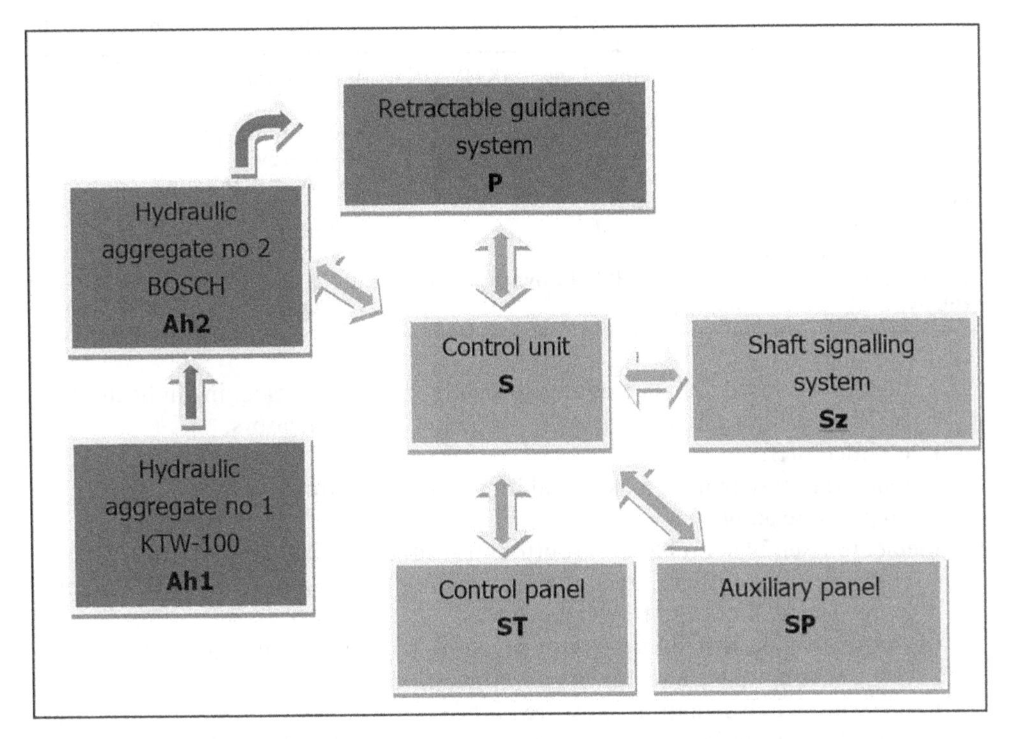

Figure 2.7 Flowchart of hydraulic power supply and control system work.

Flowchart of the hydraulic system is presented in Figure 2.7.

The power supply system consists of three hydraulic aggregates (1–3), pumping hydraulic medium to three identical blocks of hydraulic separators A, B and C, which allow to control hydraulic medium flow to actuators. Hydraulic aggregates owned by the mine were used, which allowed to eliminate potential problems with service and maintenance of aggregates. As hydraulic aggregate no. 1, the power unit of AM-50 road header was used. This device was chosen because of the ability to work in extremely difficult conditions. As no. 2 device, BOSCH hydraulic aggregate was applied, which previously was used in power supply and control system of main hoisting system break in the Leon IV shaft. It was adapted for work together with hydraulic separators. This device was used because of high degree of reliability, which is an effect of thought-out design and high quality of materials used in construction. The system was supplemented with the third aggregate with parameters identical as in case of the first aggregate. Technical parameters of hydraulic aggregates are presented in Table 2.1.

The control unit is described by functional diagram:

I. Elements of power supply unit:
   - tanks with total capacity of 400 L,
   - KTW-100 aggregate pump,
   - BOSCH aggregate pump,
   - check valves,

Table 2.1 Technical parameters of hydraulic aggregates (Dokumentacja Techniczna 2018)

Parameter	*Aggregate no. 1, no. 3 (KTW-100)*	*Aggregate no. 2 (BOSCH)*
Type of pump	Axial, multi-plunger	Radial, multi-plunger
Maximum unit yield	55 cm^3/rev	16 cm^3/rev
Electric motor	13 kW n = 1,475 rpm	7.5 kW, n = 960 rpm
Operating capacity	35 dm^3/min	16 dm^3/min at 1,000 rpm
Maximum pressure	350 bar	280 bar
Operational pressure	160 bar	160 bar
Tank capacity	200 dm^3 (HLP 46 hydraulic fluid)	200 dm^3 (HLP 46 hydraulic fluid)
Weight	650 kg	400 kg

- high pressure oil filters on aggregates with optical filter clogging indicators,
- oil level sensors with electrical high- and low-level indicators,
- oil inlet filter,
- temperature sensor with electrical limit value indicator,
- oil pressure gauge;

II. Elements assembled on block A (bottom transom stabilization):
- coil controlled distribution valve, connecting power supply unit with reciprocating space of bottom transom cylinder,
- coil controlled distribution valve, opening the return path for medium from reciprocating space of cylinders to retract guides,
- pressure switch signalling pressure in the system,
- not-related functional elements, i.e., bridging and blanking plates for separation of not-in-use paths of medium circulation (safety return path, doubled separator series-system and connection to proportional valve block);

III. Elements assembled on block B (top transom stabilization):
- coil controlled distribution valve, connecting power supply unit with reciprocating space of bottom transom cylinder,
- coil controlled distribution valve, opening the return path for medium from reciprocating space of cylinders to retract guides,
- pressure switch signalling pressure in the system,
- not-related functional elements, i.e., bridging and blanking plates for separation of not-in-use paths of medium circulation (safety return path, doubled separator series-system and connection to proportional valve block);

IV. Elements assembled on block C (guides retracting):
- coil controlled distribution valve, connecting the power supply unit with reciprocating space of bottom transom cylinder,
- coil controlled distribution valve, opening the return path for medium from reciprocating space of cylinders to retract guides,
- pressure switch signalling pressure in the system,
- not-related functional elements, i.e., bridging and blanking plates for separation of not-in-use paths of medium circulation (safety return path, doubled separator series-system and connection to proportional valve block).

Work of one or more pumps is possible to increase the medium flow. By turning on aggregate no. 1 or 3 (KTW-100), the medium is pumped through pressure leaf filter and

DN12 hydraulic pipe equipped with a check valve to the electrohydraulic system installed on aggregate no. 2. The medium is pumped to electrohydraulic system through the check valve after turning on aggregate no.2

Supply pipes of no. 1 and 2 pumps are connected before high pressure oil filter system, one filter for each hydraulic block, used to assemble separators. The oil level in the aggregate controlled by indicators and condition of oil filters are optically shown on the control device panel.

Figure 2.8 presents schemes of work of the system.

2.1.1 Bottom transom stabilizing

Operation of bottom transom stabilizing consists of movement of the bottom section of the retractable guidance system, which is done by elements assembled to the block A. After receiving a signal, they pump the medium into reciprocating space of bottom actuators both on the west and east side.

After receiving the "bottom transom stabilizing" signal, separator A moves into "b" position, allowing the cylinder to actuate. The medium returns through the separator C on block C, which opens the return path in the position.

Cylinders' actuating is stopped in the moment when the bottom section touches the bottom transom of the cage. In this moment, the target position is reached, and the

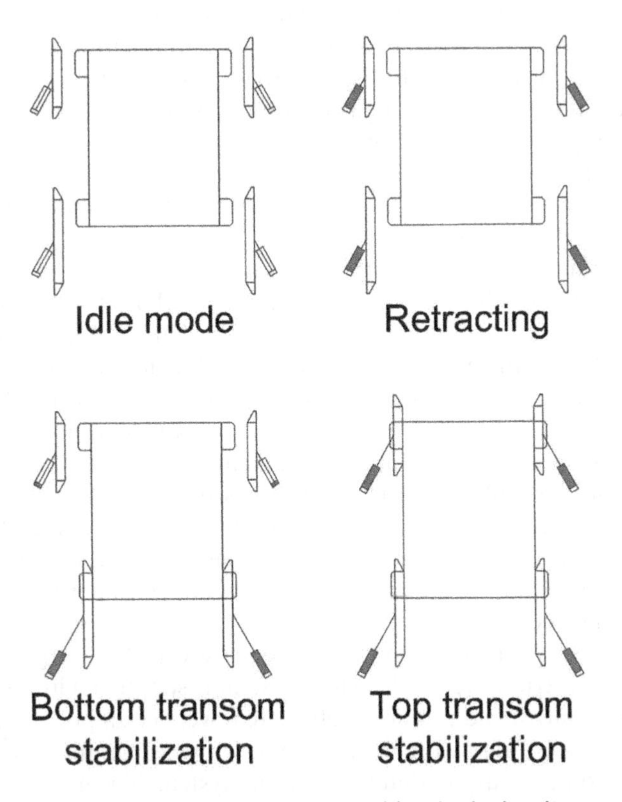

Figure 2.8 Schemes of retractable guidance control by the hydraulic system.

pressure in actuators reaches the maximum value, set by the safety valve of the pump. Separators A and C are closed when they no longer receive the "bottom transom stabilizing" signal.

2.1.2 Top transom stabilizing

Operation of top transom stabilizing consists of movement of the bottom section of the retractable guidance system, which is done by elements assembled to the block B. After receiving a signal, they pump the medium into reciprocating space of bottom actuators both on west and east side.

After receiving the "bottom transom stabilizing" signal, separator B moves into "b" position, allowing the cylinder to actuate. The medium returns through the separator C on block C, which opens the return path in the position.

Cylinders' actuating is stopped in the moment when the top section touches the bottom transom of the cage. In this moment, the target position is reached, and the pressure in actuators reaches the maximum value, set by the safety valve of the pump. Separators C and B are closed when they no longer receive the "top transom stabilizing" signal.

2.1.3 Retracting of the guides

Operation of guides retracting is done by elements assembled to the block C, which allows the medium to return. After receiving the "guides retracting" signal, the separator moves into "b" position, allowing the medium to return from the reciprocating space of actuators. The medium returns through A and B separators to A and B blocks.

After switching off the "guides retracting" signal, medium flow to actuators is stopped by the separator or, in case of already retracted guides, opens the medium return path to unload oil pressure in cylinders. Retractable guides are in idle mode, and actuators are not under pressure.

As the hydraulic control and supply system of retractable guidance system is a high-pressure installation, it is forbidden to perform work that might cause damage of the system's elements, due to safety of people in the vicinity. All activities have to be performed according to the manual, because any fault can be a reason of serious danger of hoisting system operation, which can lead to threat to people's lives and health and excessive property damage. Service operations of the hydraulic system requiring opening of system's elements has to be conducted after pressure reduction to zero and plugging the service control system. Petroleum products used as working medium can be also a threat to people's lives and health. As the oil is flammable, fire regulations must be strictly obeyed in place where the system is installed. According to existing regulations, the number of fire extinguishers at the shaft station was increased. Also, a chest with sorbent (sand) was used.

A control unit is constructed on the base of case with IP65 degree of protection, equipped with an electric system. The whole system is supplied with safe voltage 24 V. All circuits of the control and power supply system were equipped with circuit breakers. Due to low working algorithm complexity level, programmable logic devices are not included. It results in easy maintenance of the system. The majority of emergency repairs is limited to replacement of broken elements. The front of the case is made of

Figure 2.9 A view of the control device.

a transparent material and allows to control the status of the most important control signals, indicated by LED lights. A photograph of the system is presented in Figure 2.9 (Bojarski et al. 2019).

The retractable guidance system control at the level 1,000 is realized by the control panel installed at the signaller's workstation or the auxiliary control panel accessible from the conveyance. The system can be controlled using these panels when the following conditions are met:

Control panel:

- man-material cage is in the vicinity of the level 1,000,
- hoisting machine is stopped,
- signaller at the level 1,000 is allowed to control the hoisting system;

Auxiliary panel:

- man-material cage is in the vicinity of the level 1,000,
- hoisting machine is stopped,
- hoisting system in the "personal-ride" or "shaft revision" mode.

A view of the control panel is shown in Figure 2.10.

Figure 2.10 A view of the control panel.

A control signal allows also for steering of hydraulic aggregate pumps, which were selected using buttons of the panel.

A precise alignment of the cage at the level 1,000 is done by a signalman. After the hoisting machine stops, bottom transom is stabilized with the bottom section of the retractable guidance system. Monostable units were used to provide full control of the cylinders – the cylinder actuates only when the button is pushed. A position of the section is controlled with four two-state mechanical sensors. The achievement of the logical "1" by all sensors is indicated by a lamp installed on the control panel. It allows to control the top section of the retractable guidance system to stabilize the top transom. Retracting of the guides is done with one button, which causes opening of the medium return path from all actuators (Bojarski et al. 2019).

Emergency states (Dokumentacja Techniczna 2018):

A. In case of emergency state occurrence in which immediate stop of the supply and control system of retractable guides is required, the emergency stop switch installed on the control panel shall be used to switch off hydraulic pumps' power supply and control circuits.
B. In case of emergency state occurrence in which after the full work cycle guides are not fully retracted, conveyance's velocity in the vicinity of the level 1,000 is to be limited with suitable hoisting machine operation diagram to the value of speed determined in start-up operations and contained in the documentation.
C. In case of emergency or maintenance control of the retractable guidance system without the presence of the conveyance at the level 1,000, there is a possibility of remote control using the shaft signalling system by turning a key in ŁOBP port and entering the password.

2.2 CALCULATIONS

Static calculations of the retractable guidance system were carried out for its different elements, particularly top and bottom section.

2.2.1 Bottom section calculations

The following assumptions were made for the purpose of the calculations:

- Transportation unit load: 200 kN
- Distributed load:

  $$q = \frac{200\,\text{kN}}{1.654\,\text{m}} = 120.92\frac{\text{kN}}{\text{m}}$$

  Rounded value of the load is equal to 121 kN/m
- Factor of safety

  $$\frac{480}{62.87} = 7.63$$
- Load caused by hydraulic cylinder: 139.2 kN

  Data of the hydraulic unit are presented in Table 2.2.

  $$P = \frac{F}{S}$$

  $$F = P * S$$

  where

  P – hydraulic pressure, Pa;
  P = 16 MPa,
  S – area, m^2;
  $S_{piston} = 3.14 * r^2 = 0.0087\,m^2$,
  $S_{rod} = 3.14 * r^2 = 0.004\,m^2$
  $F = 16 * 10^6 * 0.0087 = 0.1392$ MN = 139.2 kN for actuating,
  $F = 16 * 10^6 * (0.0087 - 0.004) = 0.0752$ MN = 75.2 kN for actuator return.

Figures below present consecutively diagram of the retractable guidance system bottom section assembly on the level 960 m in Figure 2.11 (bottom section is coloured red), load diagram showing loads mentioned above in Figure 2.12 and model of the construction with markings of trusts, nodes and supports consecutively in Figures 2.13 and 2.14. In Figure 2.15, a diagram of the bottom section of the retractable guidance

Table 2.2 Data of the hydraulic unit

Hydraulic unit	
Multi-plunger pump:	Q = 55 cm^3/rev n = 1,400 rpm Pmax = 16 MPa
Electric engine:	Uz = 400 V Pn = 15 kW
Nominal capacity of a tank:	Qnz = 145 dm^3
Hydraulic cylinder piston	
Stroke:	1,400 mm
Piston diameter:	105 mm
Piston rod diameter:	70 mm

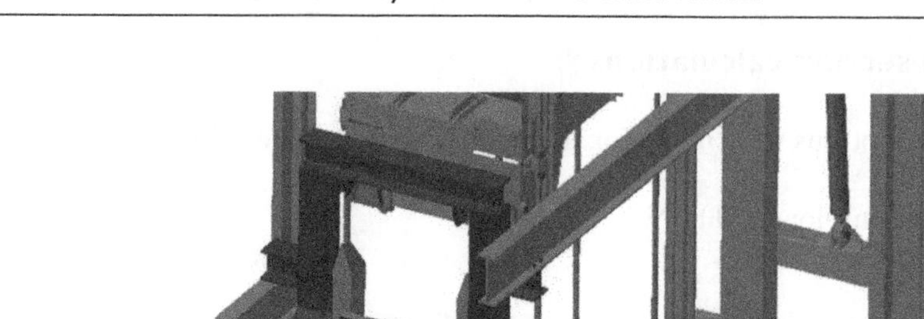

Figure 2.11 Diagram of the bottom section of the retractable guidance system assembled on the level 960.

system is shown. Figure 2.16 presents calculation data of trusts in form of a table exported from program and Figure 2.17 – table of calculation data of loads. Markings of trusts in tables match those presented in previous figures.

Figures below present models of the bottom section of the retractable guidance system. Figure 2.18 shows loads acting on the construction and Figure 2.19 presents deformations of the construction and Figure 2.20 – stress in elements of the construction.

The value of the safety factor was calculated as follows:

$$\frac{480}{62.87} = 7.63$$

2.2.2 Top section calculations

Loads acting on the top frame of the retractable guidance system are identical as in case of the bottom section. Their values are equal:

- distributed load q = 121 kN/m
- load caused by the hydraulic cylinder: 139.2 kN.

In Figure 2.21, the diagram of the construction is shown. Figures 2.22 and 2.23 present models of the top section of the retractable guidance system with supports and

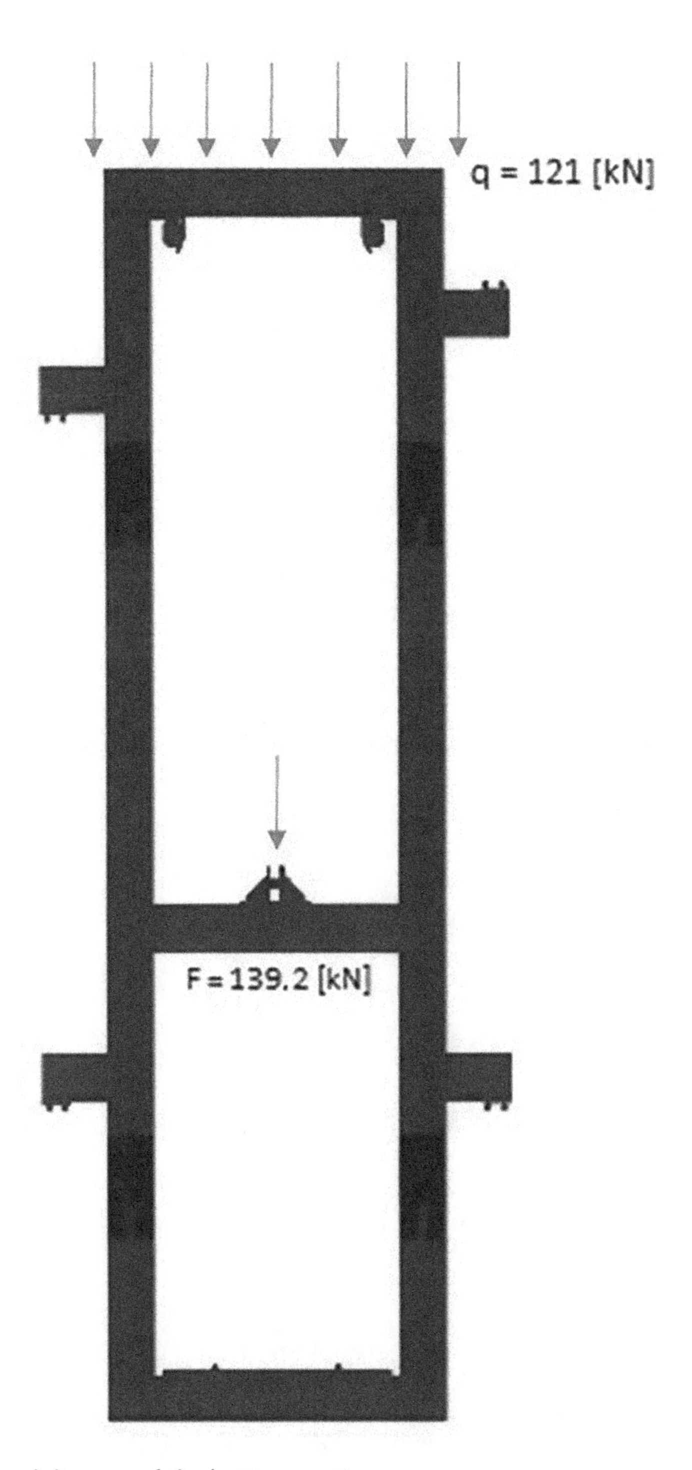

Figure 2.12 Load diagram of the bottom section.

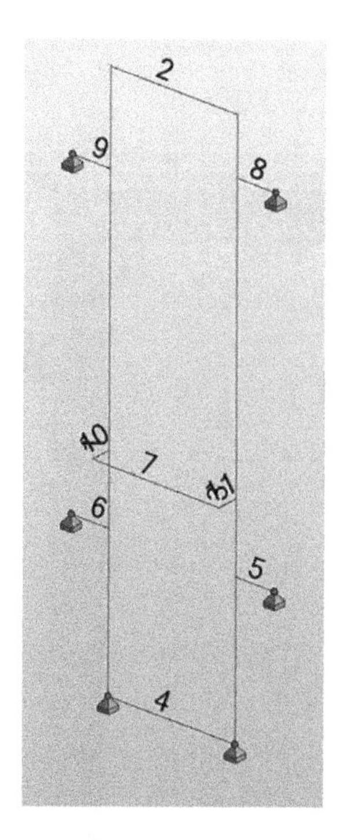

Figure 2.13 Diagram of the construction with marked trusts and supports.

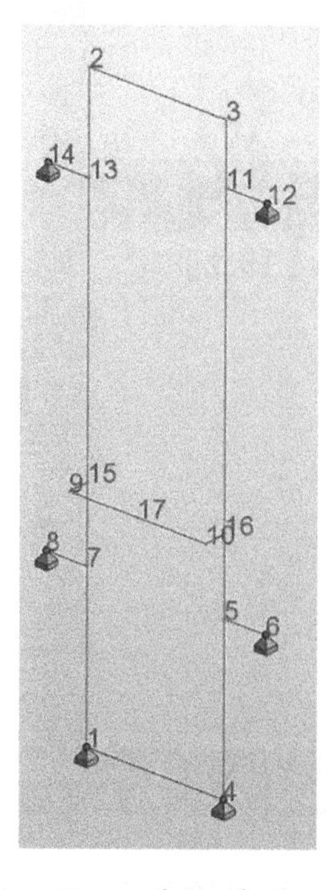

Figure 2.14 Diagram of the construction with marked nodes and supports.

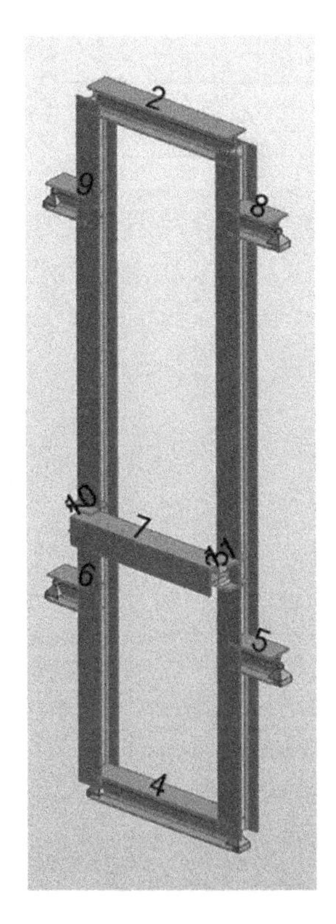

Figure 2.15 Diagram of the bottom section construction.

Pręt	Węzeł 1	Węzeł 2	Przekrój	Materiał	Gamma (Deg)	Typ	Obiekt konstrukcyjny
1	1	2	HEB 260	S 355	90,0	Belka2	Pręt
2	2	3	HEB 260	S 235	0,0	Belka2	Pręt
3	4	3	HEB 260	S 355	90,0	Belka2	Pręt
4	1	4	HEB 260	S 235	0,0	Belka2	Pręt
5	5	6	HEB 260	S 235	0,0	Belka2	Pręt
6	7	8	HEB 260	S 235	0,0	Belka2	Pręt
7	9	10	belk 260x18x2	S 355	90,0	Belka2	Pręt
8	11	12	HEB 260	S 235	0,0	Belka2	Pręt
9	13	14	HEB 260	S 235	0,0	Belka2	Pręt
10	15	9	HEB 260	S 235	0,0	Belka2	Pręt
11	16	10	HEB 260	S 235	0,0	Belka2	Pręt

Figure 2.16 Table of trusts.

	Przypadek	Typ obciążenia	Lista						
	1:STA1	ciężar własny	1do11	Cała konstruk	-Z	Wsp=1,00	MEMO :		
	2:EKSP1	siła węzłowa	17	FX=0,0	FY=0,0	FZ=-139,20	CX=0,0	CY=0,0	CZ=0,0
	2:EKSP1	obciąż. jednorodne	2	PX=0,0	PY=0,0	PZ=-121,00	globalny	nierzutowane	absolutne
*									

Figure 2.17 Table of loads.

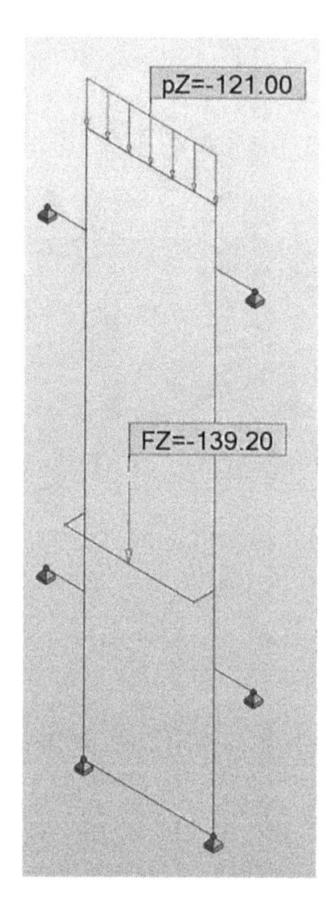

Figure 2.18 Model of loads acting on the construction.

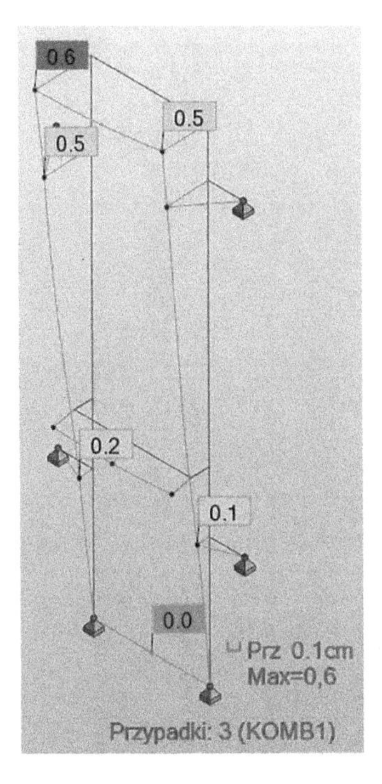

Figure 2.19 Model of deformations of the construction.

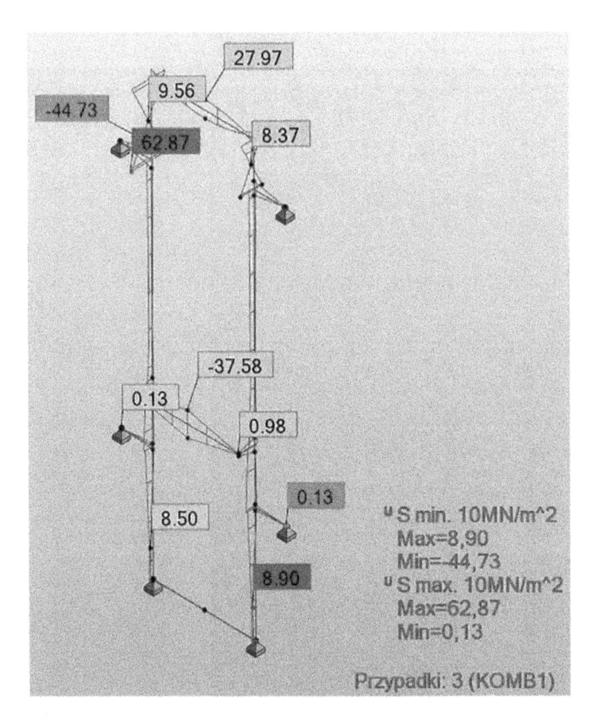

Figure 2.20 Model of stress in elements of the construction.

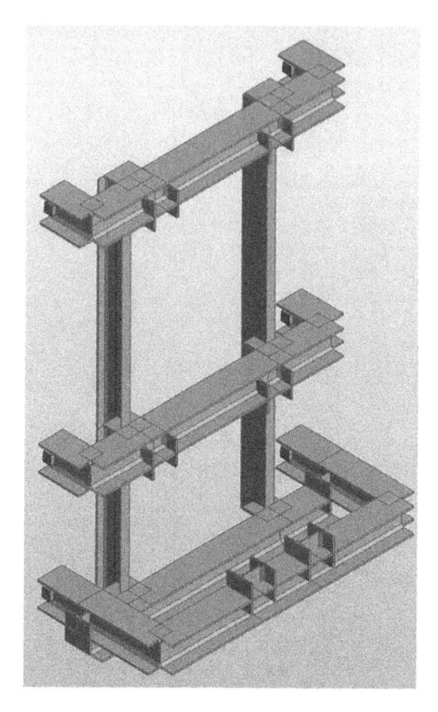

Figure 2.21 Diagram of the construction of the top section of the retractable guidance system.

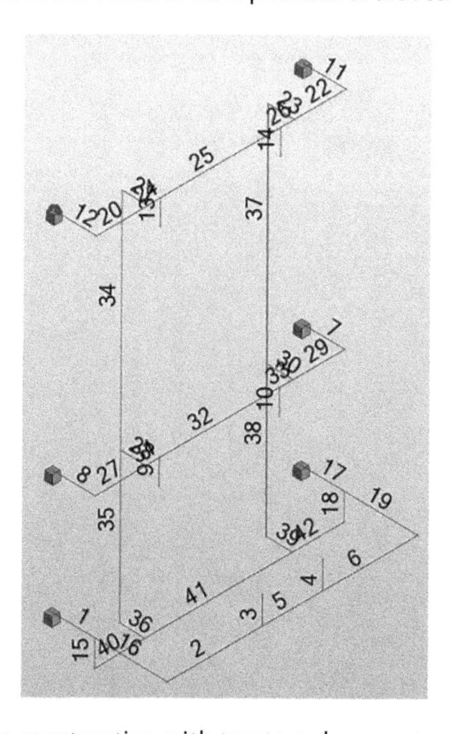

Figure 2.22 Model of the construction with trusts and supports marked.

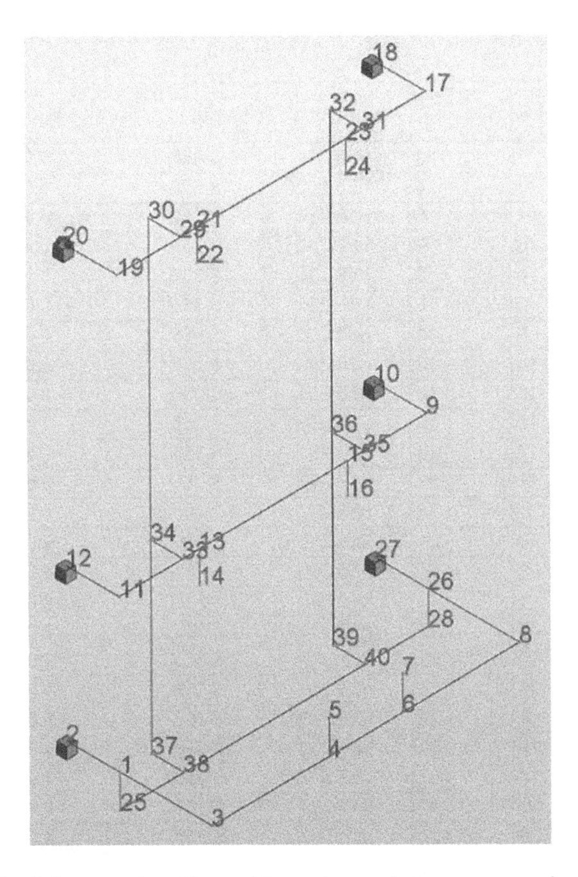

Figure 2.23 Model of the construction with nodes and supports marked.

markings of trusts (Figure 2.22) and nodes (2.23). Figures 2.24 and 2.25 present tables of consecutively calculation data of trusts and loads.

Figures below present models of the construction together with its deformations (Figure 2.26) and stress (Figure 2.27).

Similar to the bottom section, the factor of safety was calculated as follows:

$$\frac{480}{75.09} = 6.392$$

Pręty

Pręt	Węzeł 1	Węzeł 2	Przekrój	Materiał	Gamma (Deg)	Typ
1	1	2	HEB 260	S 355	0,0	Belka2
2	3	4	HEB 260	S 355	0,0	Belka2
3	4	5	HEB 260	S 355	0,0	Belka2
4	6	7	HEB 260	S 355	0,0	Belka2
5	4	6	HEB 260	S 355	0,0	Belka2
6	6	8	HEB 260	S 355	0,0	Belka2
7	9	10	HEB 260	S 355	0,0	Belka2
8	11	12	HEB 260	S 355	0,0	Belka2
9	13	14	HEB 260	S 355	0,0	Belka2
10	15	16	HEB 260	S 355	0,0	Belka2
11	17	18	HEB 260	S 355	0,0	Belka2
12	19	20	HEB 260	S 355	0,0	Belka2
13	21	22	HEB 260	S 355	0,0	Belka2
14	23	24	HEB 260	S 355	0,0	Belka2
15	1	25	HEB 260	S 355	0,0	Belka2
16	1	3	HEB 260	S 355	0,0	Belka2
17	26	27	HEB 260	S 355	0,0	Belka2
18	26	28	HEB 260	S 355	0,0	Belka2
19	26	8	HEB 260	S 355	0,0	Belka2
20	19	29	HEB 260	S 355	0,0	Belka2
21	29	30	HEB 260	S 355	0,0	Belka2
22	17	31	HEB 260	S 355	0,0	Belka2
23	31	32	HEB 260	S 355	0,0	Belka2
24	29	21	HEB 260	S 355	0,0	Belka2
25	21	23	HEB 260	S 355	0,0	Belka2
26	23	31	HEB 260	S 355	0,0	Belka2
27	11	33	HEB 260	S 355	0,0	Belka2
28	33	34	HEB 260	S 355	0,0	Belka2
29	9	35	HEB 260	S 355	0,0	Belka2
30	35	36	HEB 260	S 355	0,0	Belka2
31	33	13	HEB 260	S 355	0,0	Belka2
32	13	15	HEB 260	S 355	0,0	Belka2
33	15	35	HEB 260	S 355	0,0	Belka2
34	30	34	C 260	S 355	-90,0	Belka2
35	34	37	C 260	S 355	-90,0	Belka2
36	38	37	HEB 260	S 355	0,0	Belka2
37	32	36	C 260	S 355	90,0	Belka2
38	36	39	C 260	S 355	90,0	Belka2
39	40	39	HEB 260	S 355	0,0	Belka2
40	25	38	HEB 260	S 355	0,0	Belka2
41	38	40	HEB 260	S 355	0,0	Belka2
42	40	28	HEB 260	S 355	0,0	Belka2
*						

Figure 2.24 Table of trusts.

	Przypadek	Typ obciążenia	Lista						
	1:STA1	ciężar własny	1do42	Cała konstruk	-Z	Wsp=1,00	MEMO :		
	2:EKSP1	siła węzłowa	14 16 22 24	FX=0,0	FY=0,0	FZ=-5,00	CX=0,0	CY=0,0	CZ=0,0
	4:EKSP2	siła węzłowa	5 7	FX=0,0	FY=-38,30	FZ=-32,14	CX=0,0	CY=0,0	CZ=0,0
*									

Figure 2.25 Table of loads.

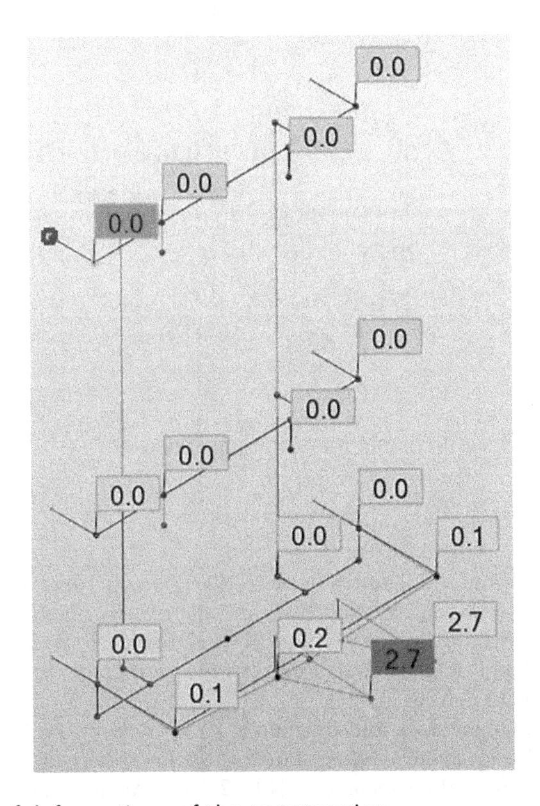

Figure 2.26 Model of deformations of the construction.

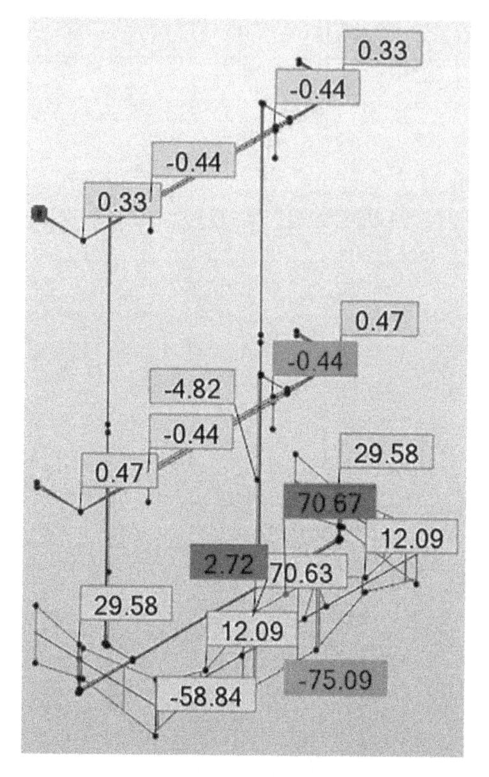

Figure 2.27 Model of stress in elements of the construction.

REFERENCES

Bojarski P, Bulenda P, Kamiński P and Nowak J. (2019) *Innowacyjny sposób stabilizacji naczyń wyciągowych wyciągu szybowego z prowadzeniem linowym na poziomie pośrednim*, in proceedings of International Mining Forum 11–12.04.2019, Katowice, Poland [In Polish].

Bulenda P. and Kamiński P. (2020) *Układ prowadzenia naczynia wyciągowego w wyrobisku szybowym*, PL433030A1 [In Polish].

Czaja P., Kamiński P., Olszewski J. and Bulenda P. (2018) *Polish experience in shaft deepening and mining shaft hoists extending on the example of the Leon shaft IV in the Rydułtowy mine //* World Mining Congress [online document]: [19–22 June], Astana, Kazakhstan.

Czekalski E. and Kubicki J. (1988) *Frontal guide used at the inter-levels of the Polish mining shaft*, PL 158153B1.

Dokumentacja Techniczna układu zasilania i sterowania urządzenia stabilizacji i prowadzenia klatki wielkogabarytowej na poziomie 1000 szybu Leon IV, opracowana przez PGG KWK ROW, 2018 r. Rydułtowy, praca niepublikowana [In Polish, unpublished].

Kamiński P. (2020) *Polish Experience in Shaft Deepening and Mining Shaft Hoists Elongation*, Mining Techniques-Past, Present and Future, InTech.

Kamiński P., Prostański D. and Dyczko A. (2021) Test of the retractable guidance system installed on the level 960 m in the Leon IV shaft in Rydułtowy Coal Mine, Poland, *IOP Conference Series: Materials Science and Engineering*. 1134 012001.

Olszewski J., Czaja P., Bulenda P. and Kamiński P. (2018) *Pogłębianie oraz wydłużanie górniczych wyciągów szybowych szybu Leon IV w kopalni KWK ROW ruch Rydułtowy*, Przegląd górniczy vol. 74, nr 8, s. 7–16 [In Polish].

Projekt techniczny prowadzenia klatki wielkogabarytowej na poz. 960m (1000m). Projekt nr Ry-157/218 opracowany przez PBSz S.A. Górnicze Biuro Projektów w 2018 r., Autorzy: B. Chomański, K. Pyrek, P. Kamiński, praca niepublikowana [In Polish, unpublished].

Watt I. (1972) Decking device for mine cages and the like conveyances, US 3800918.

Wowra D., Nowak J., Witkowski J., Izydorczyk P. and Kamiński P. (2017) *Wydłużenie Górniczego Wyciągu Szybowego – Szyb Leon IV* // In: IV PKG [online document]: IV Polski Kongres Górniczy 2017: 20–22.11.2017, Kraków [In Polish].

Chapter 3

Measurements

Measurements presented below were conducted between the July of 2018 and July of 2019. The tests covered the following issues:

- peak forces acting on stiff guides caused by the man-material cage,
- forces from man-material cage acting on retractable guides,
- mechanical and structural analysis of retractable guidance system.

Results of these measurements are presented in the following sections (Płachno 2018b).

3.1 MEASUREMENTS OF PEAK FORCES ACTING ON STIFF GUIDES FROM THE CAGE

The presented measurements were conducted to determine peak forces on stiff guides on the level 960m in the shaft Leon IV caused by the man-material cage. Results were obtained by measuring peak frontal and side forces acting on each of eight stiff guides, four on the top transom and another four on the bottom transom of the cage during moving a platform with the total load of 20 tons through a bottom level of the conveyance.

Because of technical issues, the tests were conducted in two stages. The first part, consisting of 12 single experimental cycles of pushing the platform through the cage, was conducted on the level 960m of the Leon IV shaft. Another part was performed out of the shaft after the finish of the first stage.

The essence of the first test stage was generation and record of specific measurement signal using specialized instrumentation. The signal is in mathematical relationship with forces that were to measure. The second stage of the test covered processing of gathered data by computer to obtain measured forces. The method of such processing was presented by M. Płachno in several publications (e.g., Płachno 2005). Desirable results are obtained using specialized instrumentation, which was used by Labniz in numerous tests in mine shafts. This device allows one to achieve uncertainty of measurement of the range of 12%.

A special clamping system was designed to attach a measuring device to the top and the bottom transom of the man-material cage in the Leon IV shaft. The elements of this system and the man-material cage with the measuring device attached to it are presented in Figures 3.1–3.3. The clamping system allowed the first stage of test to be conducted in the shaft in two phases:

DOI: 10.1201/b22695-3

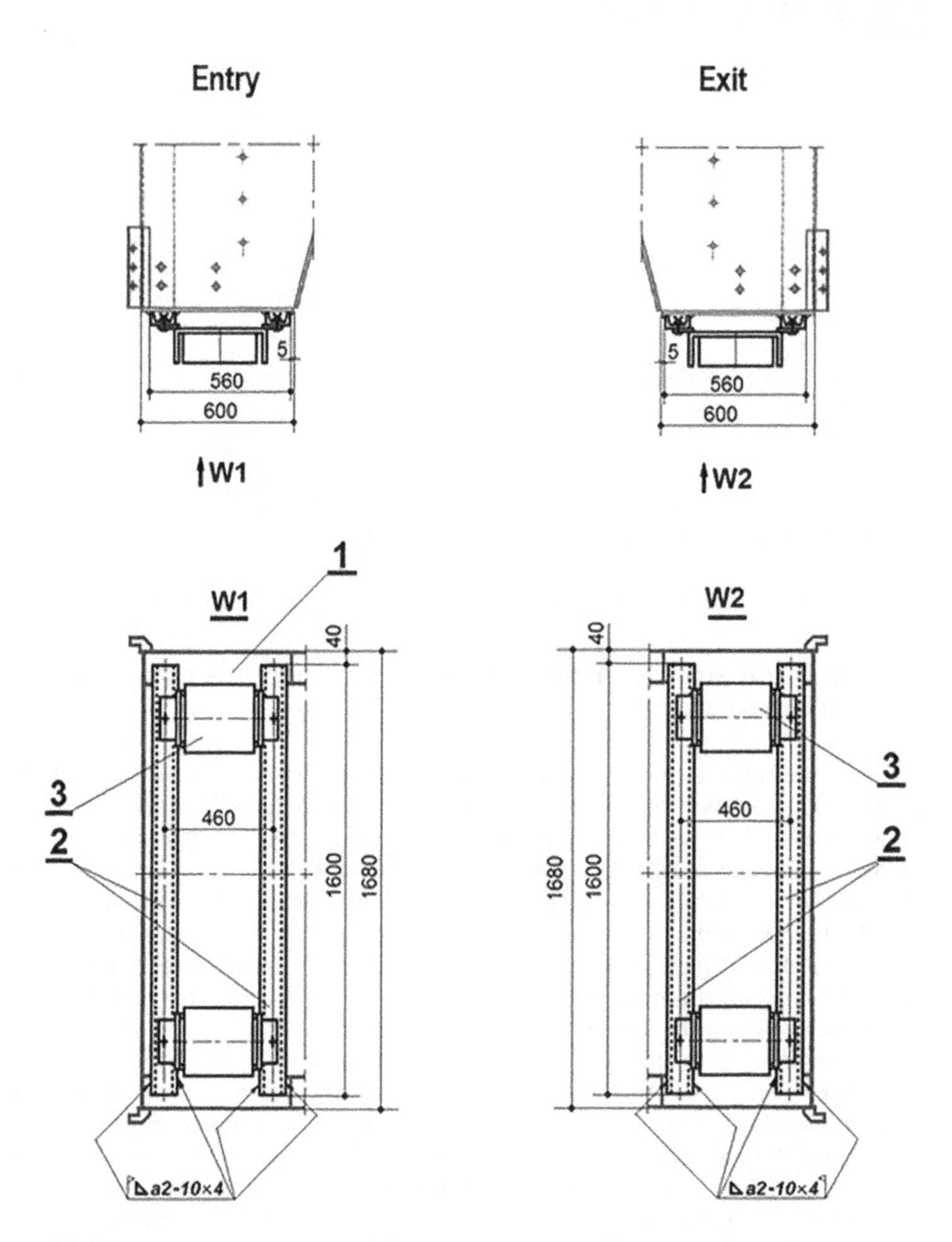

Figure 3.1 Bottom transom prepared for measurements of the horizontal forces acting during the movement of the platform through the bottom level of the man-material cage.

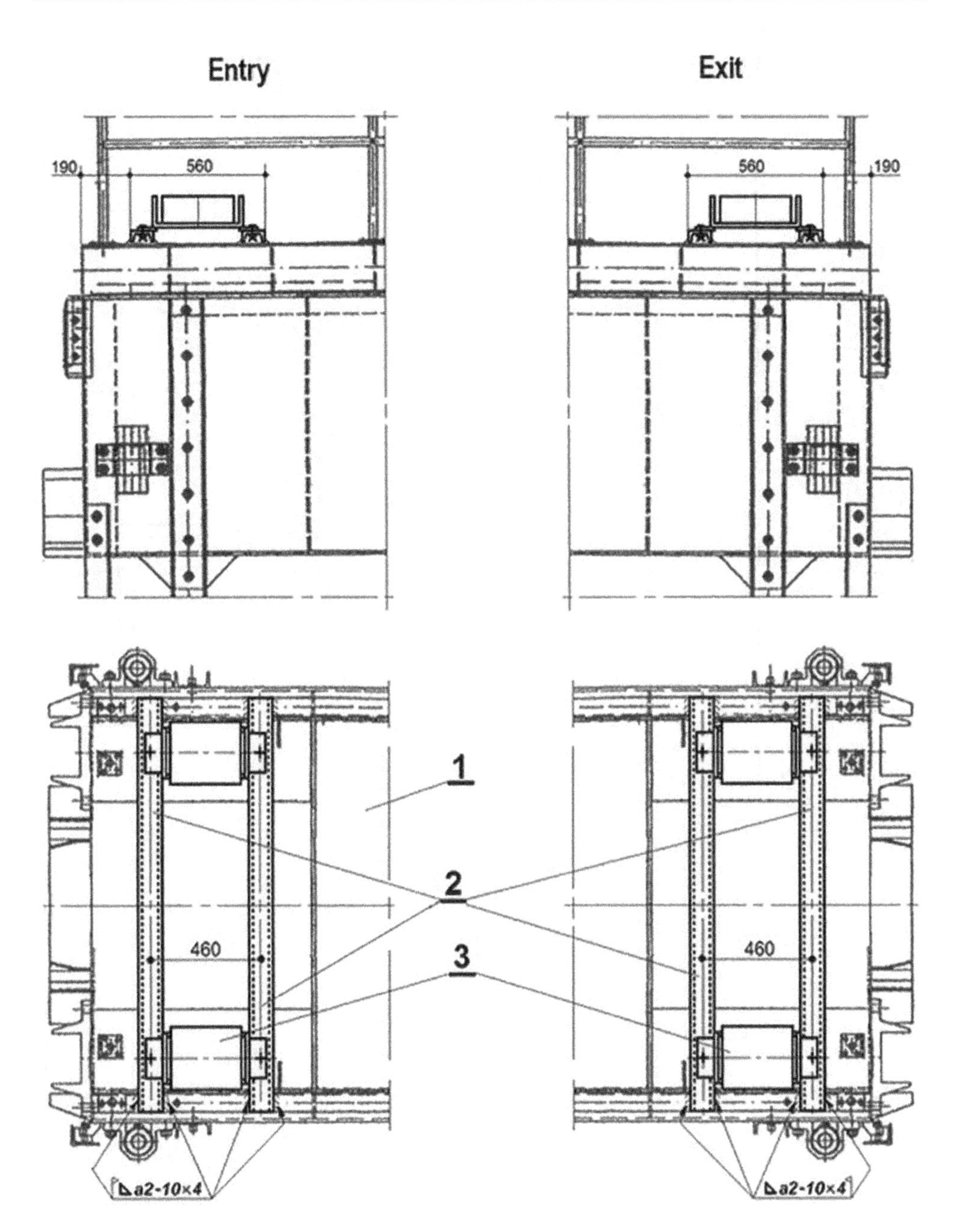

Figure 3.2 Top transom of the man-material cage prepared for measurements of the horizontal forces during movement of the platform; (a) man-material cage, (b) clamping system, (c) measuring device.

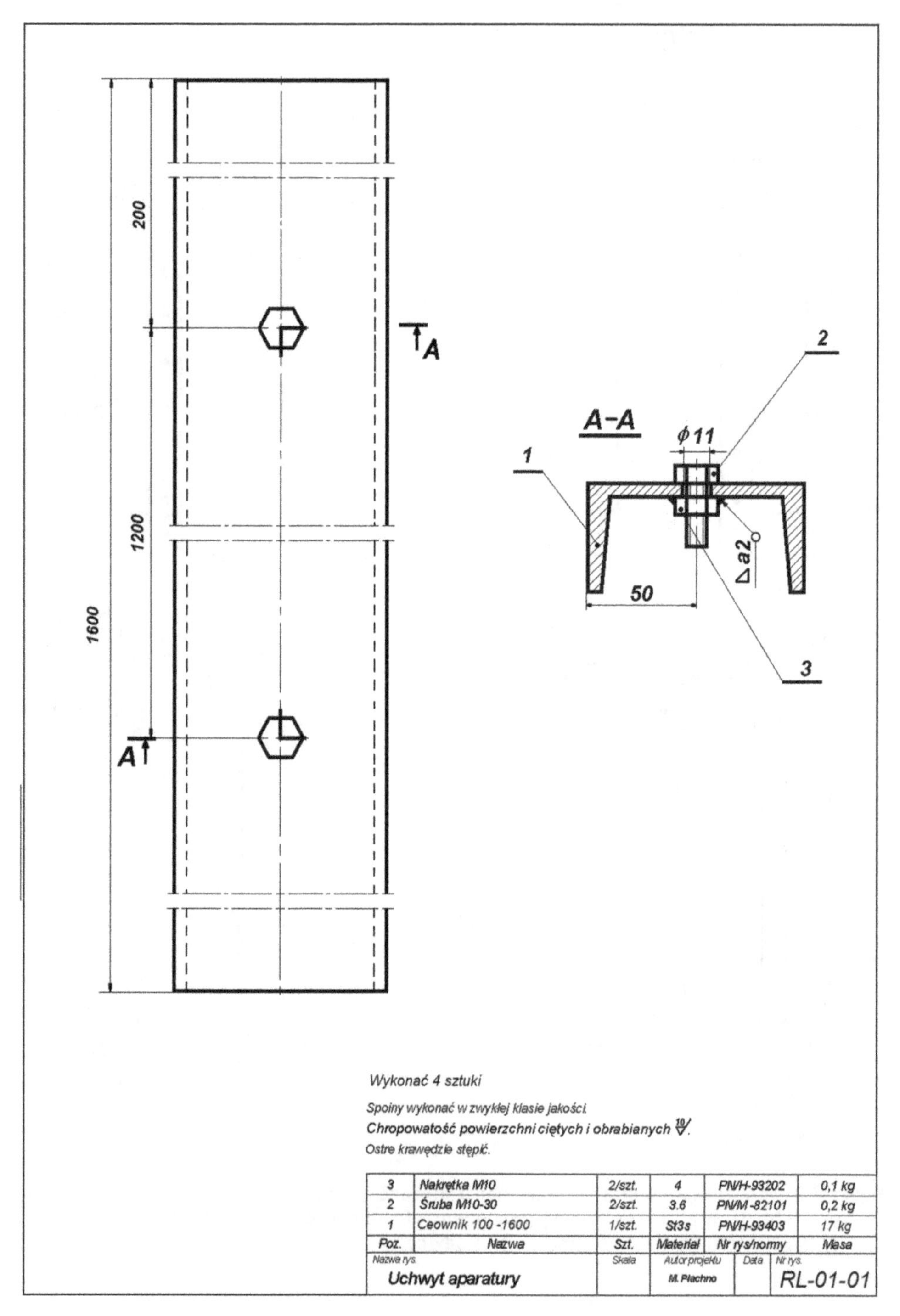

3	Nakrętka M10	2/szt.	4	PN/H-93202	0,1 kg
2	Śruba M10-30	2/szt.	3.6	PN/M-82101	0,2 kg
1	Ceownik 100 -1600	1/szt.	St3s	PN/H-93403	17 kg
Poz.	Nazwa	Szt.	Materiał	Nr rys/normy	Masa
Nazwa rys. Uchwyt aparatury		Skala	Autor projektu M. Płachno	Data	Nr rys. RL-01-01

Figure 3.3 Clamping system; (a) channel section 100–1,600, (b) hexagon screw M10-30, hexagon full nut M10.

- phase one, conducted to measure forces on the bottom transom of the man-material cage, consists of six experimental cycles of moving the platform through the bottom level of the conveyance,
- phase two, conducted to measure forces on the top transom of the man-material cage, consists of six experimental cycles of moving the platform through the bottom level of the conveyance.

Figures 3.4 and 3.5 present diagrams. The diagram presented in the Figure 3.4 presents forces measured on the bottom transom of the man-material cage, while the one shown in the Figure 3.5 presents forces measured on the top transom of the conveyance.

Markings of measured forces, presented in the figures above, consist of the following:

- numeric prefix **1**, **2**, **3** or **4**, which match first, second, third or fourth phase of the platform movement through the bottom level of the cage, where
  - phase **1** comprises the entrance of the front wheels of the platform on the cage bottom level,
  - phase **2** comprises of stabilizing the platform in the conveyance,
  - phase **3** comprises of the exit of the front wheels of the platform from the conveyance,
  - phase **4** comprises of the exit of the rear wheels of the platform from the cage.
- letter **P** in each of the marking in the Figures 3.4. and 3.5. informs that the force was measured on the bottom transom (from Polish *pomost*), while letter **G** – on the top transom (*głowica*).
- letter **N** informs if the force acted on the guide located on the **north**, while letter **S** – on the **south**.

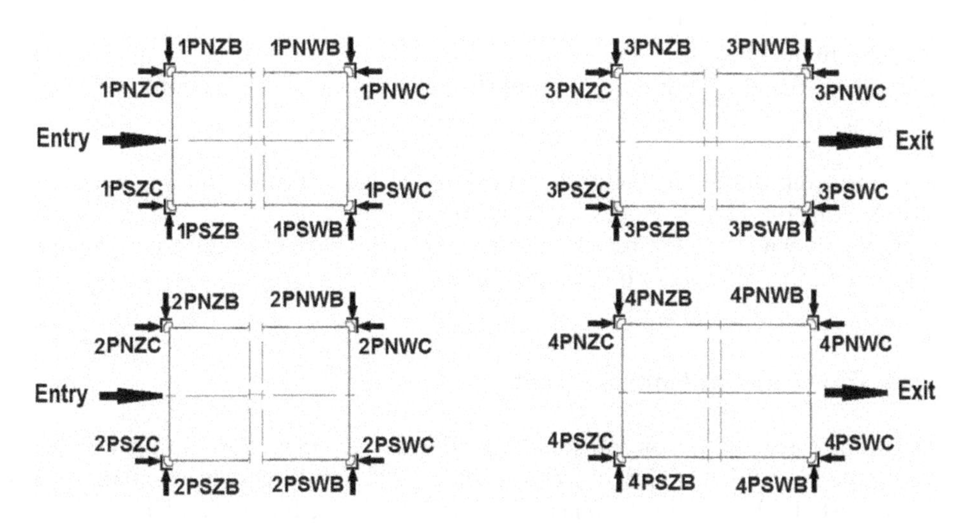

Figure 3.4 Diagram of the forces on the bottom transom of the man-material cage.

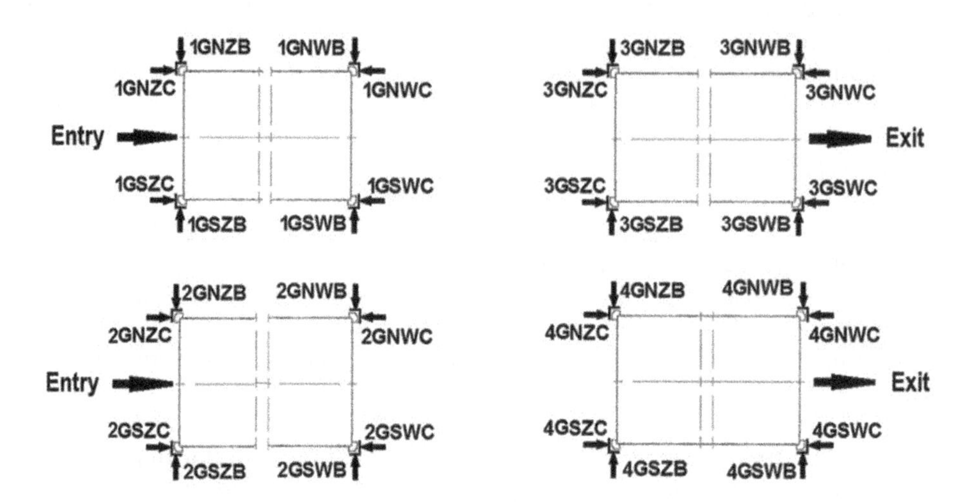

Figure 3.5 Diagram of the forces on the top transom of the man-material cage.

- letter **Z** informs that the force acted on the guide located on the side of the entry of the platform into the cage (*zapychanie*), while letter **W** – on the side of the exit from the conveyance (*wyjazd*).
- letter **C** in each marking corresponds to the frontal forces (*czołowa*) and letter **B** to the side forces (*boczna*).

3.1.1 Bottom transom frontal forces

Peak frontal forces of the bottom transom, labelled with markings shown in the Figure 3.4, are presented in Tables 3.1–3.4, of which Table 3.1 presents forces measured in the first phase of test, Table 3.2 – in the phase 2, Table 3.3 – in the phase 3 and Table 3.4 – in the phase 4.

Figures 3.6 and 3.7 present graphs of the peak forces' distribution for each phase of the test:

- graphs from the Figure 3.6 present peak forces caused by the man-material cage and acting on the stiff guides on the northern side,
- graphs shown in the Figure 3.7 present peak forces caused by the conveyance and acting on the stiff guides on southern side.

3.1.2 Top transom frontal forces

Peak frontal forces of the top transom labelled with markings shown in the Figure 3.5 are presented in Tables 3.5–3.8, of which Table 3.5 presents forces measured in the first phase of test, Table 3.6 – in the phase 2, Table 3.7 – in the phase 3 and Table 3.8 – in the phase 4.

Table 3.1 Peak frontal forces in the first phase of movement of the platform through the bottom level of the man-material cage in the first six experimental cycles

Marking of the Force	*Force value (kN)*					
	1. Cycle	*2. Cycle*	*3. Cycle*	*4. Cycle*	*5. Cycle*	*6. Cycle*
1PNWC	25.7	20.5	23.8	42.3	27.9	16.2
1PNZC	16.8	30.0	14.9	23.7	35.9	12.6
1PSWC	32.8	28.3	23.5	26.8	13.5	17.3
1PSZC	27.8	25.2	35.4	36.0	25.4	36.6

Table 3.2 Peak frontal forces in the second phase of movement of the platform through the bottom level of the man-material cage in the first six experimental cycles

Marking of the Force	*Force Value (kN)*					
	1. Cycle	*2. Cycle*	*3. Cycle*	*4. Cycle*	*5. Cycle*	*6. Cycle*
2PNWC	18.9	22.8	8.6	20.7	19.3	12.5
2PNZC	32.5	31.3	8.3	25.0	27.9	22.7
2PSWC	28.5	16.3	7.2	15.1	12.8	18.5
2PSZC	29.5	15.2	7.4	25.1	24.3	8.6

Table 3.3 Peak frontal forces in the third phase of movement of the platform through the bottom level of the man-material cage in the first six experimental cycles

Marking of the Force	*Force Value (kN)*					
	1. Cycle	*2. Cycle*	*3. Cycle*	*4. Cycle*	*5. Cycle*	*6. Cycle*
3PNWC	32.7	48.3	32.9	15.2	22.3	31.5
3PNZC	26.3	27.8	19.5	27.5	28.4	35.8
3PSWC	25.7	27.0	25.7	36.2	22.9	16.0
3PSZC	26.3	28.2	25.0	29.5	24.2	31.3

Table 3.4 Peak frontal forces in the fourth phase of movement of the platform through the bottom level of the man-material cage in the first six experimental cycles

Marking of the Force	*Force Value (kN)*					
	1. Cycle	*2. Cycle*	*3. Cycle*	*4. Cycle*	*5. Cycle*	*6. Cycle*
4PNWC	19.8	22.3	36.8	29.6	6.2	24.1
4PNZC	19.5	27.2	19.8	17.2	14.0	21.3
4PSWC	26.3	30.2	22.5	16.6	6.8	21.3
4PSZC	23.4	21.3	21.8	37.2	12.1	40.1

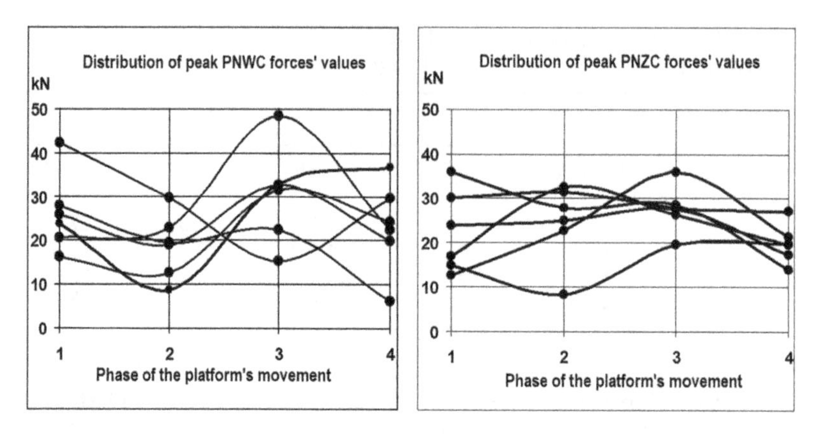

Figure 3.6 Diagrams of distributions of peak frontal forces' values acting on the northern bottom stiff guides on the level 960 m in different phases of the platform's movement through the bottom level of the cage.

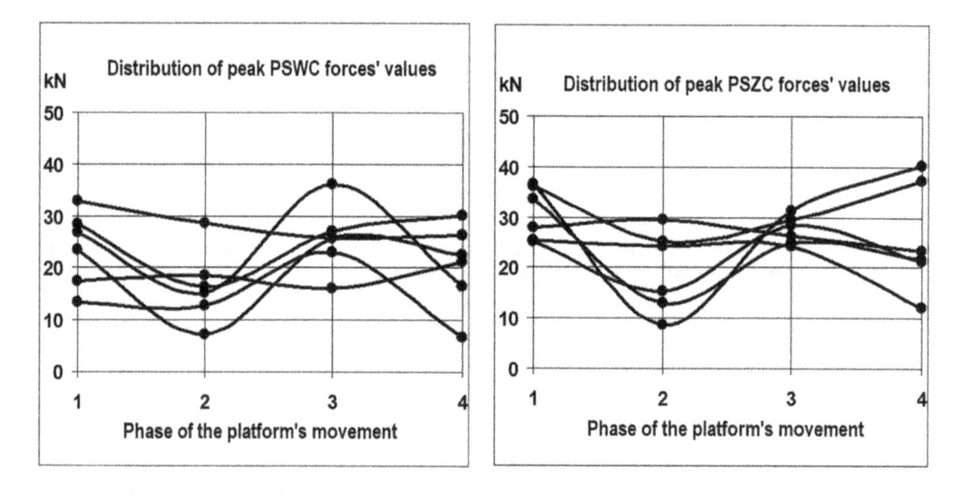

Figure 3.7 Diagrams of distributions of peak frontal forces' values acting on the southern bottom stiff guides on the level 960 m in different phases of the platform's movement through the bottom level of the cage.

Table 3.5 Peak frontal forces in the first phase of movement of the platform through the bottom level of the man-material cage in the six experimental cycles

Marking of the Force	*Force Value (kN)*					
	7. Cycle	*8. Cycle*	*9. Cycle*	*10. Cycle*	*11. Cycle*	*12. Cycle*
IGNWC	12.3	17.4	11.4	23.9	22.7	28.4
IGNZC	10.8	11.2	13.9	21.5	19.7	21.6
IGSWC	13.3	29.1	15.5	24.7	21.3	19.2
IGSZC	9.1	16.8	15.9	31.1	17.7	36.7

Table 3.6 Peak frontal forces in the second phase of movement of the platform through the bottom level of the man-material cage in the six experimental cycles

Marking of the Force	*Force Value (kN)*					
	7. Cycle	*8. Cycle*	*9. Cycle*	*10. Cycle*	*11. Cycle*	*12. Cycle*
2GNWC	12.5	20.8	8.7	16.3	22.1	9.5
2GNZC	10.3	8.6	6.5	13.1	23.8	17.4
2GSWC	15.2	16.9	10.2	17.9	17.5	23.0
2GSZC	8.9	19.8	7.7	14.9	18.1	16.1

Table 3.7 Peak frontal forces in the third phase of movement of the platform through the bottom level of the man-material cage in the six experimental cycles

Marking of the Force	*Force Value (kN)*					
	7. Cycle	*8. Cycle*	*9. Cycle*	*10. Cycle*	*11. Cycle*	*12. Cycle*
3GNWC	15.6	20.0	13.2	15.5	15.8	12.2
3GNZC	13.5	15.7	17.3	31.4	20.7	16.9
3GSWC	13.9	30.3	12.3	26.5	12.5	20.9
3GSZC	18.1	19.9	15.2	25.4	13.3	19.2

Table 3.8 Peak frontal forces in the fourth phase of movement of the platform through the bottom level of the man-material cage in the six experimental cycles

Marking of the Force	*Force Value (kN)*					
	7. Cycle	*8. Cycle*	*9. Cycle*	*10. Cycle*	*11. Cycle*	*12. Cycle*
4GNWC	15.5	13.9	7.5	6.7	8.2	44.3
4GNZC	16.1	25.6	13.2	13.5	18.5	35.9
4GSWC	13.2	20.2	7.8	9.4	19.8	48.2
4GSZC	17.6	24.0	16.1	11.5	8.4	5.1

Figures 3.8 and 3.9 present graphs of the peak forces' distribution for each phase of the test:

- graphs from Figure 3.8 present peak forces caused by the man-material cage and acting on the stiff guides on the northern side,
- graphs shown in the Figure 3.9 present peak forces caused by the conveyance and acting on the stiff guides on the southern side.

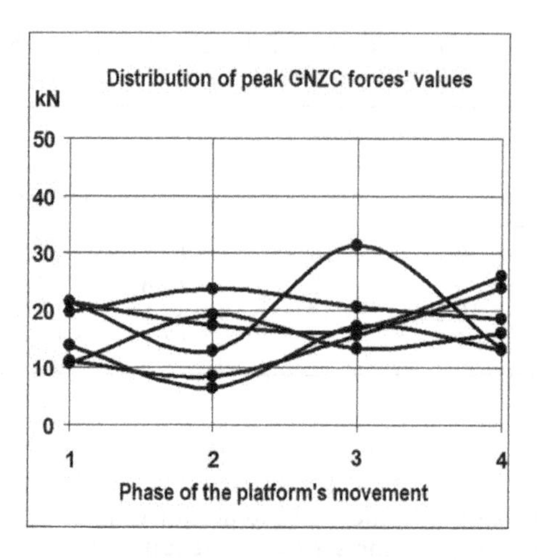

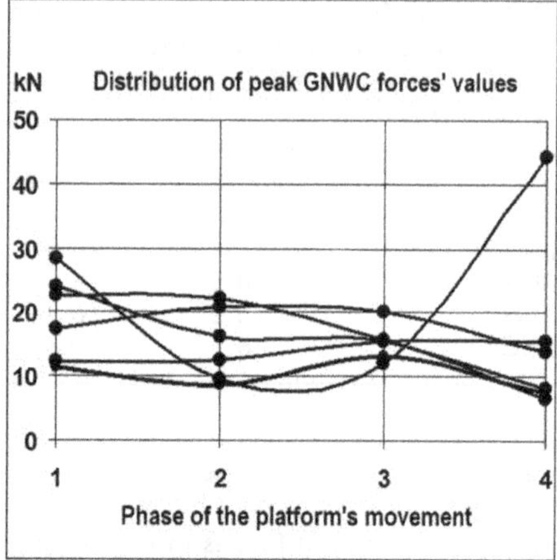

Figure 3.8 Diagrams of distributions of peak frontal forces' values acting on the northern top stiff guides on the level 960 m in different phases of the platform's movement through the bottom level of the cage.

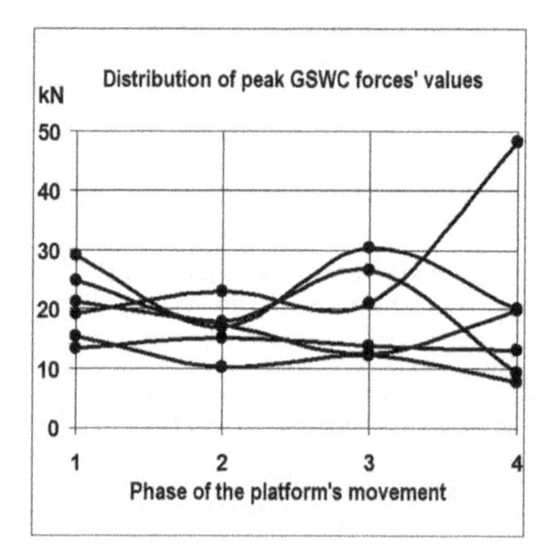

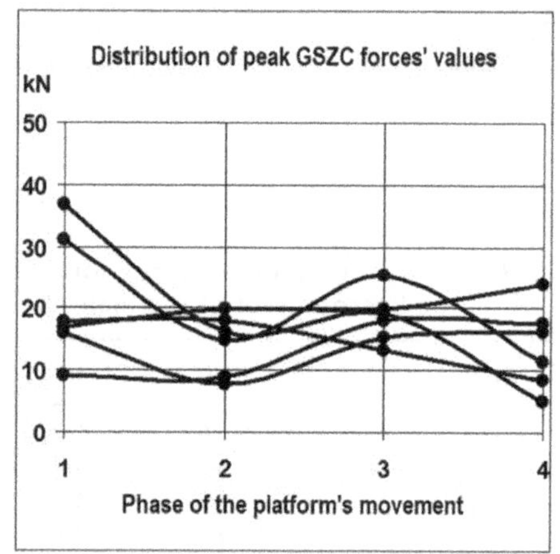

Figure 3.9 Diagrams of distributions of peak frontal forces' values acting on the southern top stiff guides on the level 960 m in different phases of the platform's movement through the bottom level of the cage.

3.1.3 Bottom transom side forces

Peak side forces of the bottom transom labelled with markings shown in the Figure 3.4 are presented in Tables 3.9–3.12, of which Table 3.9 presents forces measured in the first phase of test, Table 3.10 – in the phase 2, Table 3.11 – in the phase 3 and Table 3.12 – in the phase 4.

Table 3.9 Peak side forces in the first phase of movement of the platform through the bottom level of the man-material cage in the six experimental cycles

Marking of the Force	*Force Value (kN)*					
	1. Cycle	*2. Cycle*	*3. Cycle*	*4. Cycle*	*5. Cycle*	*6. Cycle*
1PNWB	3.7	12.9	8.3	14.1	7.3	4.4
1PNZB	7.8	12.4	15.4	8.2	7.4	11.2
1PSWB	8.9	15.1	16.5	11.0	7.8	13.8
1PSZB	19.8	12.0	13.7	14.5	13.2	12.1

Table 3.10 Peak side forces in the second phase of movement of the platform through the bottom level of the man-material cage in the six experimental cycles

Marking of the Force	*Force Value (kN)*					
	1. Cycle	*2. Cycle*	*3. Cycle*	*4. Cycle*	*5. Cycle*	*6. Cycle*
2PNWB	4.8	5.9	4.5	5.3	2.4	1.5
2PNZB	5.8	6.6	7.6	7.5	1.8	2.1
2PSWB	6.4	6.3	4.9	3.1	2.2	2.6
2PSZB	5.8	5.2	5.4	6.3	1.8	2.5

Table 3.11 Peak side forces in the third phase of movement of the platform through the bottom level of the man-material cage in the six experimental cycles

Marking of the Force	*Force Value (kN)*					
	1. Cycle	*2. Cycle*	*3. Cycle*	*4. Cycle*	*5. Cycle*	*6. Cycle*
3PNWB	4.9	4.8	12.8	13.1	9.1	8.2
3PNZB	7.5	8.8	6.2	5.1	8.7	8.9
3PSWB	5.3	5.4	16.7	15.9	8.3	5.8
3PSZB	7.0	12.3	4.2	4.0	17.3	14.5

Table 3.12 Peak side forces in the fourth phase of movement of the platform through the bottom level of the man-material cage in the six experimental cycles

Marking of the Force	*Force Value (kN)*					
	1. Cycle	*2. Cycle*	*3. Cycle*	*4. Cycle*	*5. Cycle*	*6. Cycle*
4PNWB	18.0	13.5	20.1	7.9	11.2	11.9
4PNZB	4.9	5.3	3.8	3.6	8.9	3.9
4PSWB	8.5	11.5	8.5	17.8	12.0	10.2
4PSZB	6.9	3.2	4.9	3.6	11.8	6.4

Figures 3.10 and 3.11 present graphs of the peak forces' distribution for each phase of the test:

- graphs from Figure 3.10 present peak forces caused by the man-material cage and acting on the stiff guides on the northern side,
- graphs shown in Figure 3.11 present peak forces caused by the conveyance and acting on the stiff guides on southern side.

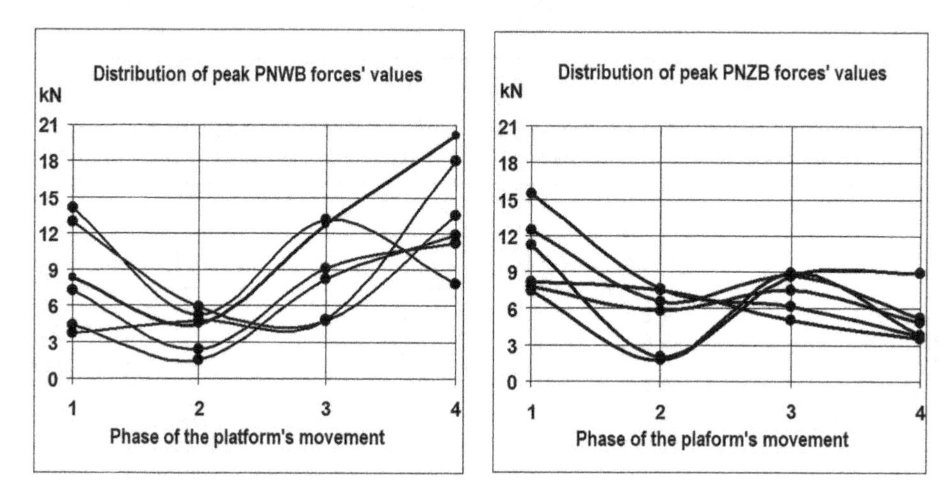

Figure 3.10 Diagrams of distributions of peak side forces' values acting on the northern bottom stiff guides on the level 960 m in different phases of the platform's movement through the bottom level of the cage.

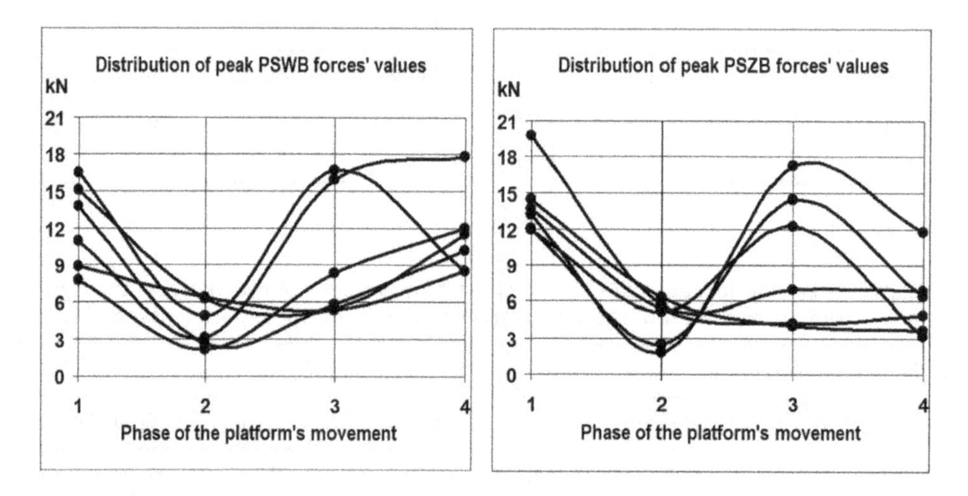

Figure 3.11 Diagrams of distributions of peak side forces' values acting on the southern bottom stiff guides on the level 960 m in different phases of the platform's movement through the bottom level of the cage.

3.1.4 Top transom side forces

Peak side forces of the top transom, labelled with markings shown in Figure 3.5 are presented in Tables 3.13–3.16, of which Table 3.13 presents forces measured in the first phase of test, Table 3.14 – in the phase 2, Table 3.15 – in the phase 3 and Table 3.16 – in the phase 4.

Figures 3.12 and 3.13 present graphs of the peak forces' distribution for each phase of the test:

- graphs from the Figure 3.12 present peak forces caused by the man-material cage and acting on the stiff guides on the northern side,
- graphs shown in Figure 3.13 present peak forces caused by the conveyance and acting on the stiff guides on the southern side.

Table 3.13 Peak side forces in the first phase of movement of the platform through the bottom level of the man-material cage in the six experimental cycles

Marking of the Force	*Force Value (kN)*					
	7. Cycle	*8. Cycle*	*9. Cycle*	*10. Cycle*	*11. Cycle*	*12. Cycle*
1GNWB	3.0	4.4	2.6	3.9	5.5	2.0
1GNZB	4.3	5.5	3.4	12.5	3.3	7.5
1GSWB	7.5	5.8	3.5	3.2	6.5	4.9
1GSZB	4.4	5.1	7.4	6.2	3.2	9.5

Table 3.14 Peak side forces in the second phase of movement of the platform through the bottom level of the man-material cage in the six experimental cycles

Marking of the Force	*Force Value (kN)*					
	7. Cycle	*8. Cycle*	*9. Cycle*	*10. Cycle*	*11. Cycle*	*12. Cycle*
2GNWB	3.4	2.8	2.4	3.6	5.1	3.1
2GNZB	2.1	2.5	2.2	2.4	1.6	3.3
2GSWB	2.7	3.7	2.2	2.1	3.6	2.5
2GSZB	3.6	1.5	2.1	7.5	1.9	2.4

Table 3.15 Peak side forces in the third phase of movement of the platform through the bottom level of the man-material cage in the six experimental cycles

Marking of the Force	*Force Value (kN)*					
	7. Cycle	*8. Cycle*	*9. Cycle*	*10. Cycle*	*11. Cycle*	*12. Cycle*
3GNWB	3.3	7.4	4.4	4.2	3.1	3.5
3GNZB	3.1	3.5	4.0	3.5	6.1	5.4
3GSWB	3.9	6.2	3.2	2.0	3.6	2.5
3GSZB	2.7	3.1	5.1	6.7	5.8	5.2

Table 3.16 Peak side forces in the third phase of movement of the platform through the bottom level of the man-material cage in the six experimental cycles

Marking of the Force	*Force Value (kN)*					
	7. Cycle	*8. Cycle*	*9. Cycle*	*10. Cycle*	*11. Cycle*	*12. Cycle*
4GNWB	2.3	3.7	4.3	11.5	9.1	8.2
4GNZB	4.2	3.5	8.0	6.7	3.3	2.8
4GSWB	2.8	3.0	1.9	4.1	6.4	18.4
4GSZB	2.5	4.3	7.3	7.0	4.3	3.5

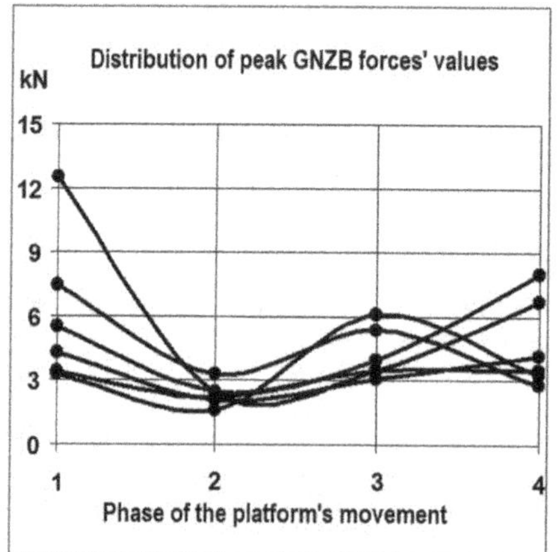

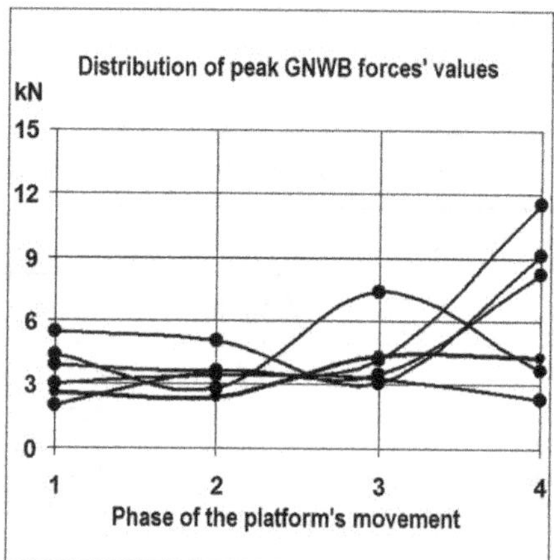

Figure 3.12 Diagrams of distributions of peak side forces' values acting on the northern top stiff guides on the level 960 m in different phases of the platform's movement through the bottom level of the cage.

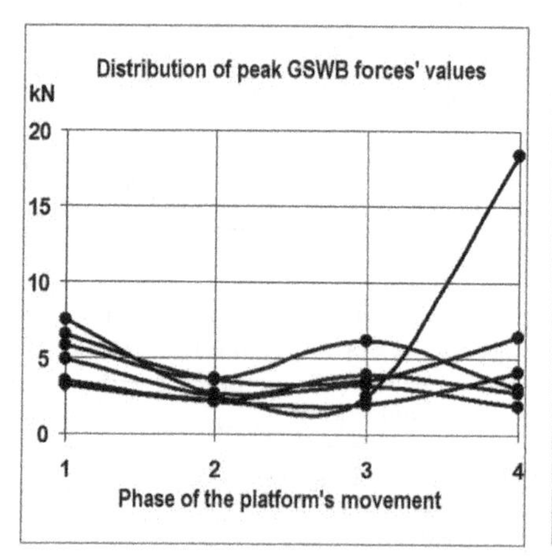

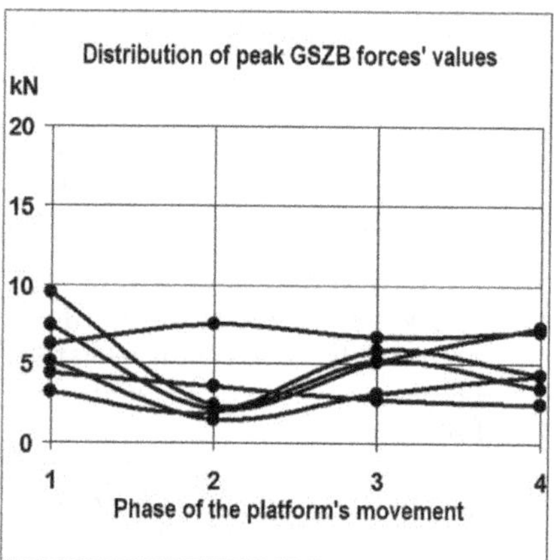

Figure 3.13 Diagrams of distributions of peak side forces' values acting on the southern top stiff guides on the level 960 m in different phases of the platform's movement through the bottom level of the cage.

3.1.5 Conclusions

Tables 3.17 and 3.18 present selected peak forces of measured forces, eight for each stiff guide on the top and bottom transom. Frontal forces are shown in Table 3.17 and side forces in Table 3.18.

On the basis of tests' results presented above, the following conclusions were formulated:

1. Values of frontal forces acting on the stiff guidance on the level 960 m and caused by the man-material cage during loading and unloading its bottom level are diverse, because of clearances between guides and the cage as well as random dynamics of

   a. Entry of the platform's front and rear wheels into the cage's level,
   b. Exit of the platform's front and rear wheels from the conveyance's level.

   Dynamic's randomness was recognized as the main reason of the occurrence of the peak frontal force equal to 42.3 kN during loading and 48.3 kN during unloading of the conveyance.

Table 3.17 Peak frontal forces caused by the man-material cage acting on the stiff guidance

Force Marking	*Force Value (kN)*							
	1. Phase		*2. Phase*		*3. Phase*		*4. Phase*	
	Bottom Transom	*Top Transom*	*Bottom Transom*	*Top Transom*	*Bottom Transom*	*Top Transom*	*Bottom Transom*	*Top Transom*
NWC	42.3	28.4	29.7	22.1	48.3	20.0	36.8	44.3
NZC	35.9	21.6	32.5	23.8	35.8	31.4	27.2	25.9
SWC	32.8	29.1	28.5	23.0	36.2	30.3	30.3	48.2
SZC	36.6	36.7	29.5	19.8	31.3	25.4	40.1	24.0

Table 3.18 Peak side forces caused by the man-material cage acting on the stiff guidance

Force Marking	*Force value (kN)*							
	1. Phase		*2. Phase*		*3. Phase*		*4. Phase*	
	Bottom Transom	*Top Transom*	*Bottom Transom*	*Top Transom*	*Bottom Transom*	*Top Transom*	*Bottom Transom*	*Top Transom*
NWB	14.1	5.5	5.9	5.1	13.1	7.4	20.1	11.5
NZB	15.4	12.5	7.6	**3.3**	8.9	6.1	8.9	8.0
SWB	16.5	7.5	6.4	3.7	16.7	6.2	17.8	18.4
SZB	19.8	9.5	5.8	7.5	17.3	6.7	11.8	7.3

2. Results obtained suggest that peak frontal forces acting on the stiff guides and caused by cage's guide shoe on the top transom should be considered equal to peak forces caused by the guide shoe located near the bottom transom of the cage.
3. Measured peak side forces did not exceed the level of 42% of peak frontal forces. It should be taken into account that the biggest values of side forces occur in the case of derailing of the platform during its movement through the conveyance, which did not happen during the test.

3.2 MEASUREMENTS OF PEAK FORCES ACTING ON THE RETRACTABLE GUIDES

The test aimed at determining the following:

- peak forces acting on the retractable guidance system on the level 960m in the Leon IV shaft during loading and unloading of the man-material cage,
- peak stress in the bridle hangers during loading and unloading of the conveyance.

Because of the technical conditions, the test was parted into two stages. First of them was conducted in the Leon IV shaft, and it comprised of generation and record of measuring signals in 12 experimental cycles of moving the platform with the load of 20 tons through the bottom level of the man-material cage. Measuring signals are in a relationship with measured forces and stress. The second stage was carried out after the finish of the first part, and it consists of processing of obtained signals. It resulted in values of forces and stress and graphs of their distribution.

Presented tests were conducted similar to previous measurements conducted for the case of stiff guidance on the level 960m and presented in the previous section. Retractable guidance system stabilizes the conveyance on the level of its top and bottom transom, and it does not provide additional stabilization in the middle of the cage (on the floor of second cage level). Such feature was not considered necessary, which statement was needed to verify. Because of that, a range of measurements were extended to provide values of stress in the bridle hangers.

The extension of test was realized by utilization of two additional measuring units, located on the floor of the second cage level, together with the basic measuring units, described in the previous section. All of the measuring units are synchronized, and thus, the uncertainty of measurement is the same as in the previous case, so not higher than 12%.

3.2.1 Measurement schedule

The schedule of the test was similar to the one conducted for stiff guides, presented in the previous section. It was divided into two parts:

- stage "**bottom transom**", consisting of two series, including three experimental cycles of moving the platform with a load of 20 tons through the bottom level of the cage each. During the first part, the measuring unit was located on the bottom transom of the conveyance;

- stage "**top transom**", consisting of two series, including three experimental cycles of moving the platform with a load of 20 tons through the bottom level of the cage each. During the second part, the measuring unit was located on the top transom of the conveyance.

In both stages of the experiment, additional measuring units, located on the middle beam of the cage (floor of the second cage level), were used.

3.2.2 "Bottom transom" stage

First part of the test covered the following steps:

1. First alignment of the platform with a load of 20 tons on the level 960 m, standing by to push through the bottom level of the man-material cage.
2. First alignment of the bottom level of the conveyance on the level 1,050 m for the purpose of attachment of the clamping device (Figure 3.14) to the cage, using fire dampers.
3. First assembling of the measurement unit on the man-material cage, according to Figure 3.15.
4. First alignment of the cage on the level 960 m with its bottom level ready for moving the platform through.
5. Remote opening of the first file of measurement record and first push of the platform into the bottom level of the conveyance.
6. Remote opening of the second file of measurement record and first push of the platform out of the cage. Closing of the record file.
7. First technological break of the "bottom transom" stage and second alignment of the platform on standby to another experimental cycle.
8. Remote opening of the third file of measurement record and second push of the platform into the bottom level of the conveyance.
9. Remote opening of the fourth file of measurement record and second push of the platform out of the cage. Closing of the record file.
10. Second technological break of the "bottom transom" stage and third alignment of the platform on standby to another experimental cycle.
11. Remote opening of the fifth file of measurement record and third push of the platform into the bottom level of the conveyance.
12. Remote opening of the sixth file of measurement record and third push of the platform out of the cage. Closing of the record file.
13. Third technological break:
    a. Fourth alignment of the platform on standby to another experimental cycle.
    b. Second alignment of the empty man-material cage on the level 1,050 m and attachment of the measuring unit according to Figure 3.16 using fire dampers.
14. Second alignment of the man-material cage on the level 960 m with its bottom level ready for moving the platform through.
15. Remote opening of the seventh file of measurement record and fourth push of the platform into the bottom level of the conveyance.
16. Remote opening of the eighth file of measurement record and fourth push of the platform out of the cage. Closing of the record file.

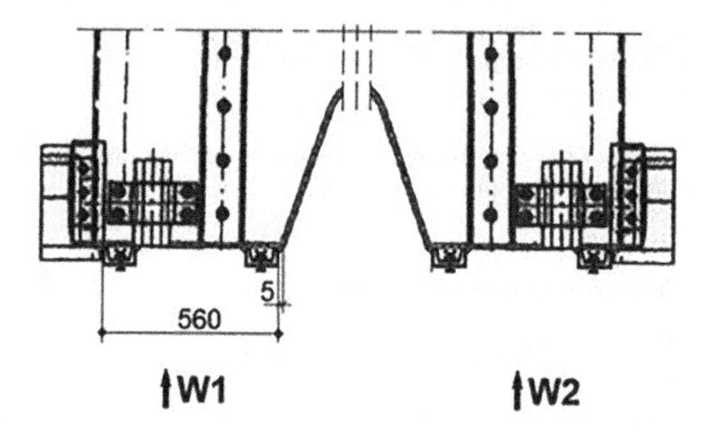

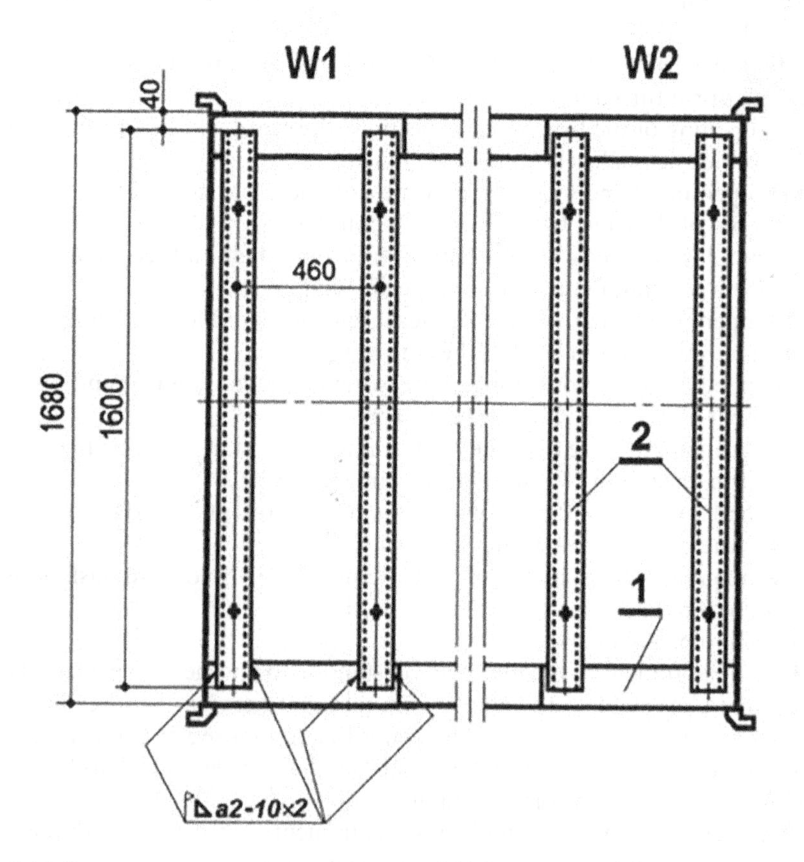

Figure 3.14 Bottom transom of the man-material cage prepared for measurements of the horizontal forces acting on the retractable guidance system on level 960 m in the Leon IV shaft; (a) man-material cage, (b) clamping system.

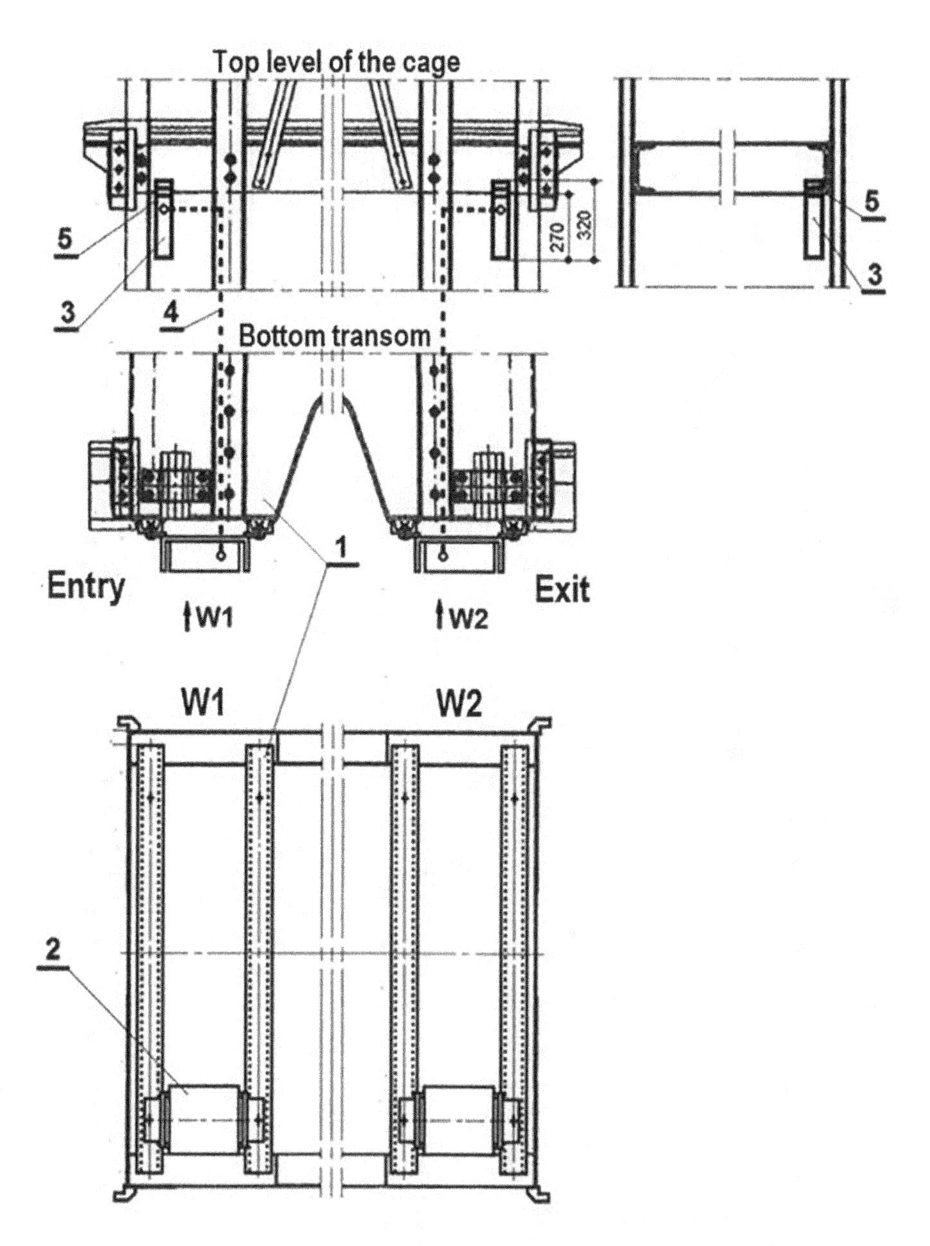

Figure 3.15 Assembly drawing of the measuring unit for the first part of measurements covering three experimental cycles of the platform's movement through the bottom level of the cage; (a) bottom transom of the man-material cage, (b) basic measuring device, (b) external measuring unit, (c) conductor cable, (d) wedge washer.

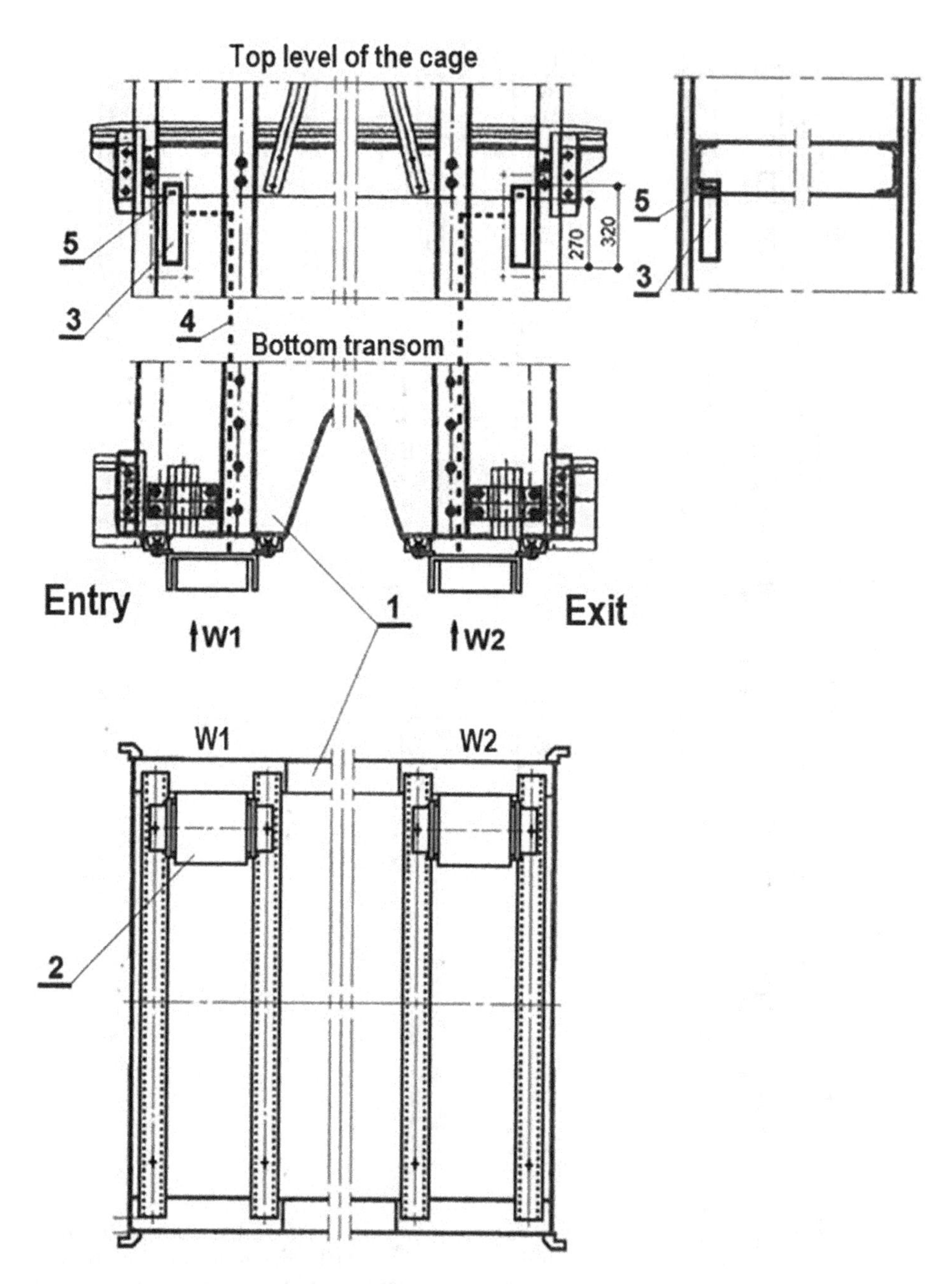

Figure 3.16 Assembly drawing of the measuring unit for the second part of measurements covering three experimental cycles of the platform's movement through the bottom level of the cage; (a) bottom transom of the man-material cage, (b) basic measuring device, (c) external measuring unit, (d) conductor cable, (e) wedge washer.

17. Fourth technological break of the "bottom transom" stage and fifth alignment of the platform on standby to another experimental cycle.
18. Remote opening of the ninth file of measurement record and fifth push of the platform into the bottom level of the conveyance.
19. Remote opening of the tenth file of measurement record and fifth push of the platform out of the cage. Closing of the record file.
20. Fifth technological break of the "bottom transom" stage and sixth alignment of the platform on standby to another experimental cycle.
21. Remote opening of the eleventh file of measurement record and sixth push of the platform into the bottom level of the conveyance.
22. Remote opening of the twelfth file of measurement record and sixth push of the platform out of the cage. Closing of the record file.
23. Third alignment of the bottom level of the conveyance on the level 1,050m for the purpose of disassembling of the clamping device with measuring unit using fire dampers.
24. Sixth technological break and finish of the "bottom transom" stage.

3.2.3 "Top transom" stage

1. Seventh alignment of the platform with a load of 20 tons on the level of 960m, standing by to push through the bottom level of the man-material cage.
2. First alignment of the top transom of the conveyance on the level 960m for the purpose of attachment of the clamping device (Figure 3.17) to the cage.
3. First assembling of the measurement unit on the man-material cage, according to Figure 3.18.
4. Seventh alignment of the cage on the level of 960m with its bottom level ready for moving the platform through.
5. Remote opening of the thirteenth file of measurement record and seventh push of the platform into the bottom level of the conveyance.
6. Remote opening of the fourteenth file of measurement record and seventh push of the platform out of the cage. Closing of the record file.
7. First technological break of the "top transom" stage and eighth alignment of the platform on standby to another experimental cycle.
8. Remote opening of the fifteenth file of measurement record and eighth push of the platform into the bottom level of the conveyance.
9. Remote opening of the sixteenth file of measurement record and eighth push of the platform out of the cage. Closing of the record file.
10. Second technological break of the "top transom" stage and ninth alignment of the platform on standby to another experimental cycle.
11. Remote opening of the seventeenth file of measurement record and ninth push of the platform into the bottom level of the conveyance.
12. Remote opening of the eighteenth file of measurement record and ninth push of the platform out of the cage. Closing of the record file.
13. Third technological break:

    a. Tenth alignment of the platform on standby to another experimental cycle.
    b. Second alignment of the empty man-material cage on the level 960m and attachment of the measuring unit according to Figure 3.19.

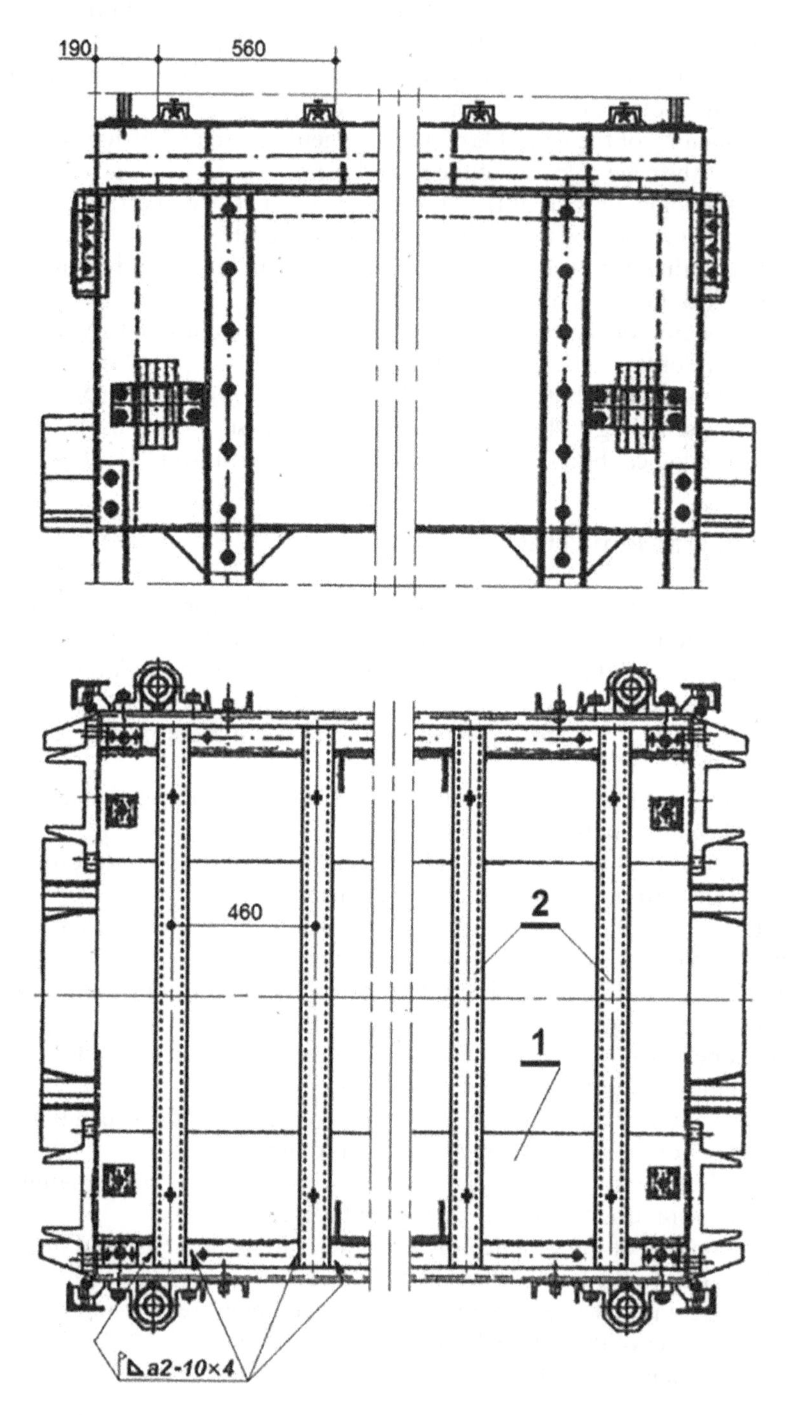

Figure 3.17 Top transom of the man-material cage prepared for measurements of the horizontal forces acting on the retractable guidance system on level 960 m in the Leon IV shaft; (a) man-material cage, (b) clamping system.

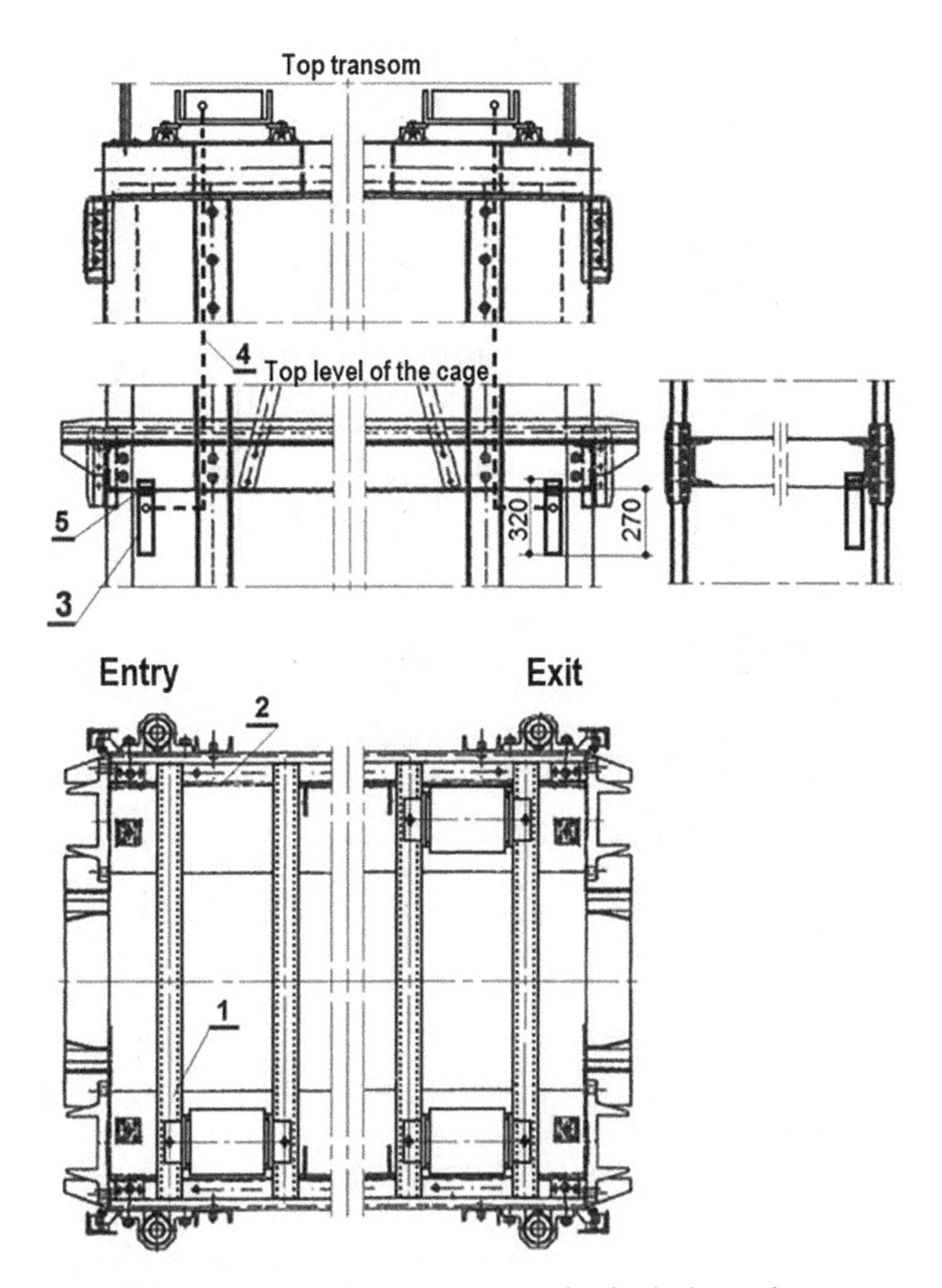

Figure 3.18 Assembly drawing of the measuring unit for the third part of measurements covering three experimental cycles of the platform's movement through the bottom level of the cage; (a) bottom transom of the man-material cage, (b) basic measuring device, (c) external measuring unit, (d) conductor cable, (e) wedge washer.

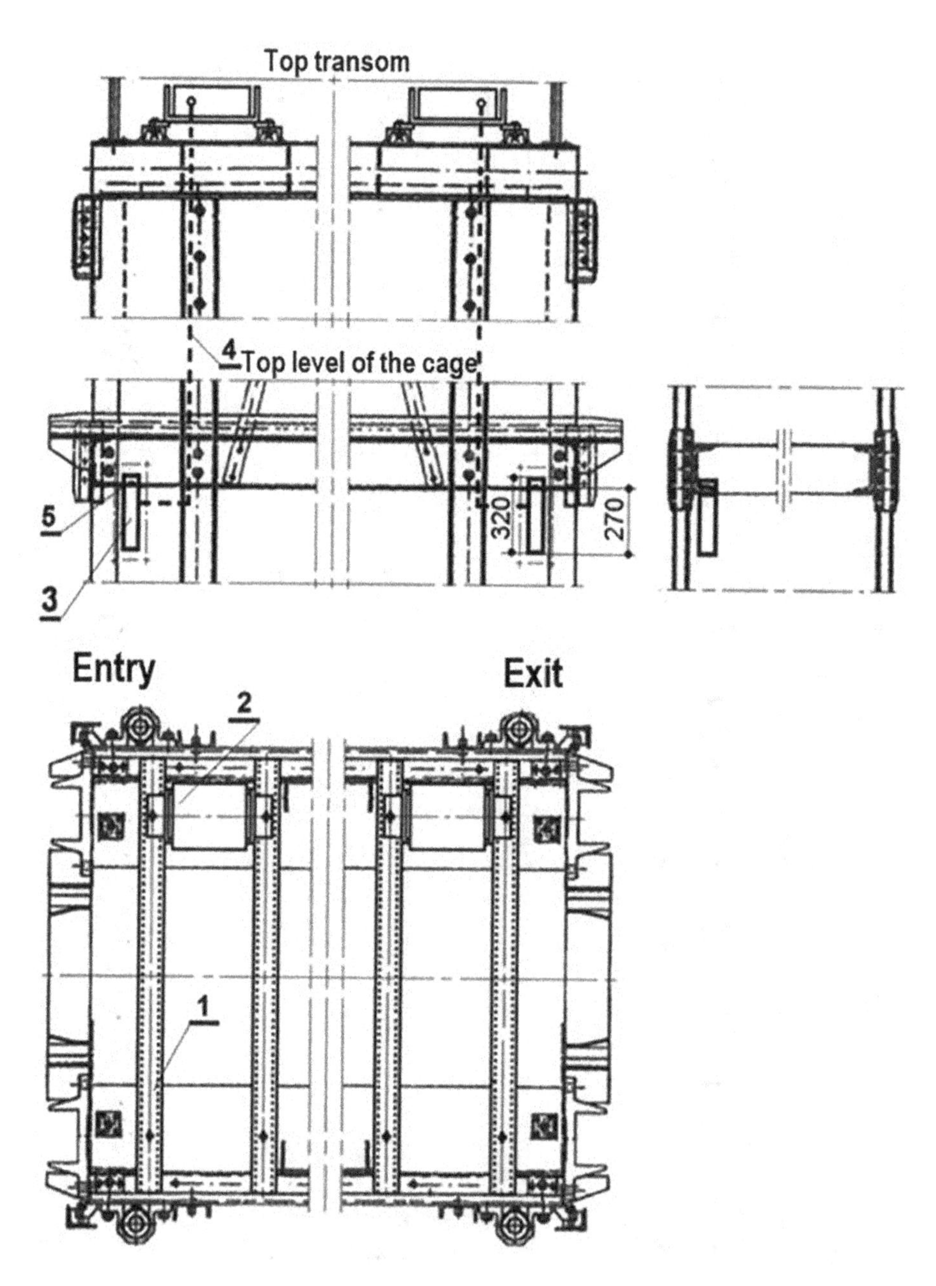

Figure 3.19 Assembly drawing of the measuring unit for the fourth part of measurements covering three experimental cycles of the platform's movement through the bottom level of the cage; (a) bottom transom of the man-material cage, (b) basic measuring device, (c) external measuring unit, (d) conductor cable, (e) wedge washer.

14. Tenth alignment of the man-material cage on the level 960 m with its bottom level ready for moving the platform through.
15. Remote opening of the nineteenth file of measurement record and tenth push of the platform into the bottom level of the conveyance.
16. Remote opening of the twentieth file of measurement record and tenth push of the platform out of the cage. Closing of the record file.
17. Fourth technological break of the "top transom" stage and eleventh alignment of the platform on standby to another experimental cycle.
18. Remote opening of the twenty-first file of measurement record and eleventh push of the platform into the bottom level of the conveyance.
19. Remote opening of the twenty-second file of measurement record and eleventh push of the platform out of the cage. Closing of the record file.
20. Fifth technological break of the "top transom" stage and twelfth alignment of the platform on standby to another experimental cycle.
21. Remote opening of the twenty-third file of measurement record and twelfth push of the platform into the bottom level of the conveyance.
22. Remote opening of the twenty-fourth file of measurement record and twelfth push of the platform out of the cage. Closing of the record file.
23. Third alignment of the top transom of the conveyance on the level 960 m for the purpose of disassembling of the clamping device with the measuring unit.
24. Finish of the measurements carried out in the Leon IV shaft.

3.2.4 Diagrams and markings of measured forces

Markings of the forces acting on the retractable guidance system and caused by the man-material cage are similar as for the case of stiff guidance. Diagrams of these forces are presented in Figures 3.4 and 3.5. Their markings consist of the following:

- numeric prefix corresponding to the phase of the platform's movement through the cage,
- letter P for bottom transom or G for top transom,
- letter N for northern guides or S for southern guides,
- letter Z for the entry side or W for the exit side,
- letter C for frontal forces or B for side forces.

3.2.5 Bottom transom frontal forces

Peak frontal forces of the bottom transom, labelled with markings shown in Figure 3.4 are presented in Tables 3.19–3.22, of which Table 3.19 presents forces measured in the first phase of test, Table 3.20 – in the phase 2, Table 3.21 – in the phase 3 and Table 3. 22 – in the phase 4.

Figures 3.20 and 3.21 present graphs of the peak forces' distribution for each phase of the test:

- graphs from Figure 3.20 present peak forces caused by the man-material cage and acting on the stiff guides on the northern side,
- graphs shown in Figure 3.21 present peak forces caused by the conveyance and acting on the stiff guides on the southern side.

Table 3.19 Peak frontal forces in the first phase of movement of the platform through the bottom level of the man-material cage in the six experimental cycles

Marking of the Force	*Force Value (kN)*					
	1. Cycle	*2. Cycle*	*3. Cycle*	*4. Cycle*	*5. Cycle*	*6. Cycle*
1PNWC	11.3	11.8	9.8	10.7	20.0	22.4
1PNZC	27.8	24.8	23.0	21.1	21.4	16.6
1PSWC	14.7	16.1	21.0	24.5	13.5	11.1
1PSZC	4.9	16.2	11.2	6.8	20.2	12.3

Table 3.20 Peak frontal forces in the second phase of movement of the platform through the bottom level of the man-material cage in the six experimental cycles

Marking of the Force	*Force Value (kN)*					
	1. Cycle	*2. Cycle*	*3. Cycle*	*4. Cycle*	*5. Cycle*	*6. Cycle*
2PNWC	21.1	24.1	18.3	21.9	18.1	21.2
2PNZC	15.5	19.0	24.1	14.3	19.4	20.9
2PSWC	25.3	21.2	23.3	26.3	14.8	18.8
2PSZC	24.9	18.6	22.2	20.8	16.5	24.5

Table 3.21 Peak frontal forces in the third phase of movement of the platform through the bottom level of the man-material cage in the six experimental cycles

Marking of the Force	*Force Value (kN)*					
	1. Cycle	*2. Cycle*	*3. Cycle*	*4. Cycle*	*5. Cycle*	*6. Cycle*
3PNWC	29.2	30.0	22.9	24.8	23.1	24.3
3PNZC	28.9	31.2	29.5	32.0	32.1	27.5
3PSWC	20.3	18.5	21.3	15.2	19.5	14.3
3PSZC	24.3	28.1	19.9	15.2	20.8	17.3

Table 3.22 Peak frontal forces in the fourth phase of movement of the platform through the bottom level of the man-material cage in the six experimental cycles

Marking of the Force	*Force Value (kN)*					
	1. Cycle	*2. Cycle*	*3. Cycle*	*4. Cycle*	*5. Cycle*	*6. Cycle*
4PNWC	38.9	30.2	36.6	34.8	29.9	38.1
4PNZC	33.2	38.3	40.8	34.5	36.4	32.1
4PSWC	34.7	35.3	40.7	39.2	38.9	33.2
4PSZC	31.5	35.9	38.3	32.9	34.2	29.5

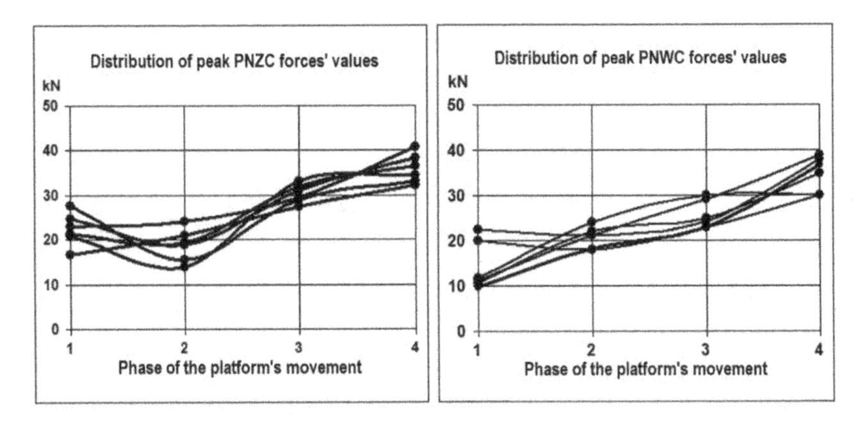

Figure 3.20 Diagrams of distributions of peak frontal forces' values acting on the northern bottom retractable guides on the level 960 m in different phases of the platform's movement through the bottom level of the cage.

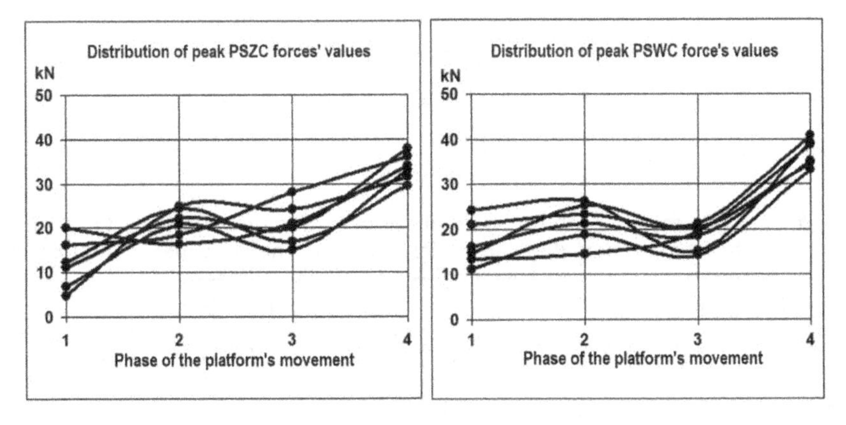

Figure 3.21 Diagrams of distributions of peak frontal forces' values acting on the southern bottom retractable guides on the level 960 m in different phases of the platform's movement through the bottom level of the cage.

3.2.6 Top transom frontal forces

Peak frontal forces of the top transom, labelled with markings shown in Figure 3.5, are presented in Tables 3.23–3.26, of which Table 3.23 presents forces measured in the first phase of test, Table 3.24 – in the phase 2, Table 3.25 – in the phase 3 and Table 3.26 – in the phase 4.

Figures 3.22 and 3.23 present graphs of the peak forces' distribution for each phase of the test:

- graphs from Figure 3.22 present peak forces caused by the man-material cage and acting on the stiff guides on the northern side,

Table 3.23 Peak frontal forces in the first phase of movement of the platform through the bottom level of the man-material cage in the six experimental cycles

Marking of the Force	*Force Value (kN)*					
	7. Cycle	*8. Cycle*	*9. Cycle*	*10. Cycle*	*11. Cycle*	*12. Cycle*
1GNWC	10.3	11.8	12.2	18.1	11.9	17.8
1GNZC	11.1	11.9	13.1	17.8	19.6	19.1
1GSWC	16.1	12.3	11.9	13.5	18.5	12.9
1GSZC	17.3	9.1	15.4	19.2	16.3	14.9

Table 3.24 Peak frontal forces in the second phase of movement of the platform through the bottom level of the man-material cage in the six experimental cycles

Marking of the Force	*Force Value (kN)*					
	7. Cycle	*8. Cycle*	*9. Cycle*	*10. Cycle*	*11. Cycle*	*12. Cycle*
2GNWC	11.2	9.8	13.5	15.5	13.7	17.9
2GNZC	15.9	17.5	16.9	13.1	18.9	19.3
2GSWC	23.4	18.1	29.5	22.5	21.2	26.1
2GSZC	16.3	16.1	23.5	19.8	18.1	12.9

Table 3.25 Peak frontal forces in the third phase of movement of the platform through the bottom level of the man-material cage in the six experimental cycles

Marking of the Force	*Force Value (kN)*					
	7. Cycle	*8. Cycle*	*9. Cycle*	*10. Cycle*	*11. Cycle*	*12. Cycle*
3GNWC	21.7	24.4	27.6	17.9	17.0	18.2
3GNZC	18.2	18.6	17.5	16.4	24.3	18.0
3GSWC	14.7	17.5	16.2	19.3	18.5	15.7
3GSZC	10.8	17.4	12.5	18.3	19.8	16.9

Table 3.26 Peak frontal forces in the fourth phase of movement of the platform through the bottom level of the man-material cage in the six experimental cycles

Marking of the Force	*Force Value (kN)*					
	7. Cycle	*8. Cycle*	*9. Cycle*	*10. Cycle*	*11. Cycle*	*12. Cycle*
4GNWC	37.3	34.2	32.9	39.7	33.1	32.3
4GNZC	34.9	31.5	28.2	29.1	32.5	27.1
4GSWC	35.3	37.7	39.2	32.9	40.2	33.3
4GSZC	28.3	35.2	29.8	32.1	36.3	34.5

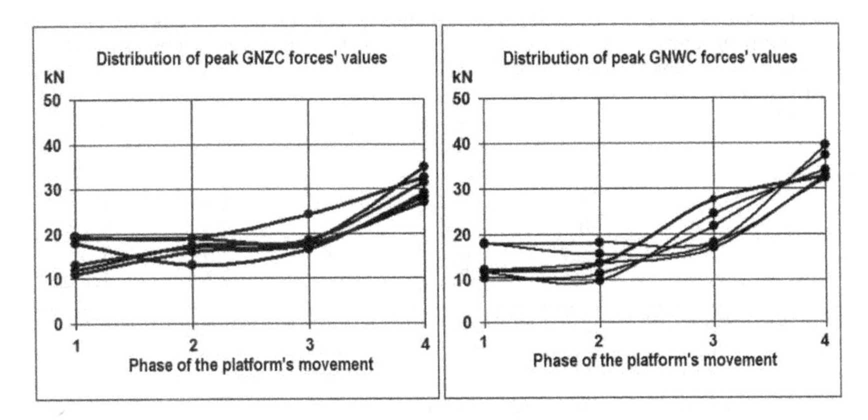

Figure 3.22 Diagrams of distributions of peak frontal forces' values acting on the northern top retractable guides on the level 960 m in different phases of the platform's movement through the bottom level of the cage.

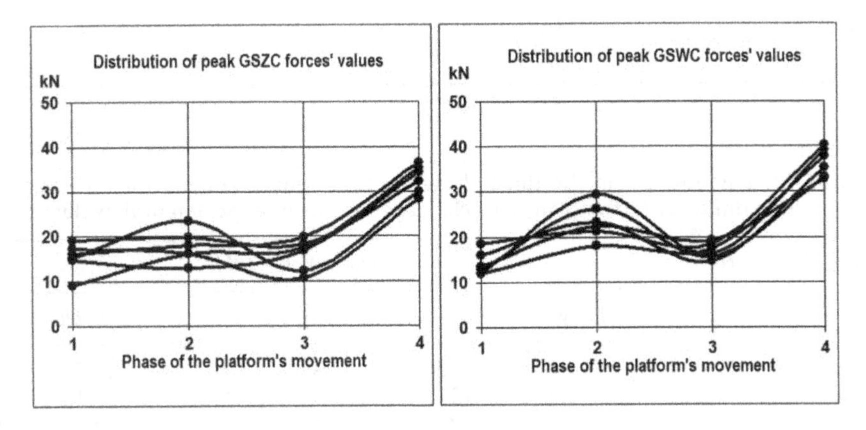

Figure 3.23 Diagrams of distributions of peak frontal forces' values acting on the southern top retractable guides on the level 960 m in different phases of the platform's movement through the bottom level of the cage.

- graphs shown in Figure 3.23 present peak forces caused by the conveyance and acting on the stiff guides on the southern side.

3.2.7 Bottom transom side forces

Peak side forces of the bottom transom labelled with markings shown in Figure 3.4 are presented in Tables 3.27–3.30, of which Table 3.27 presents forces measured in the first phase of test, Table 3.28 – in the phase 2, Table 3.29 – in the phase 3 and Table 3.30 – in the phase 4.

Table 3.27 Peak side forces in the first phase of movement of the platform through the bottom level of the man-material cage in the six experimental cycles

Marking of the Force	*Force Value (kN)*					
	1. Cycle	*2. Cycle*	*3. Cycle*	*4. Cycle*	*5. Cycle*	*6. Cycle*
1PNWB	6.5	8.3	9.7	5.7	7.0	7.3
1PNZB	8.5	6.0	5.3	7.5	5.9	7.2
1PSWB	4.9	10.2	5.4	7.9	8.8	6.4
1PSZB	3.5	3.2	6.1	7.6	4.5	3.9

Table 3.28 Peak side forces in the second phase of movement of the platform through the bottom level of the man-material cage in the six experimental cycles

Marking of the Force	*Force Value (kN)*					
	1. Cycle	*2. Cycle*	*3. Cycle*	*4. Cycle*	*5. Cycle*	*6. Cycle*
2PNWB	8.4	9.2	9.3	9.7	8.5	9.3
2PNZB	7.0	4.5	7.3	4.1	5.9	5.7
2PSWB	9.6	8.9	11.2	10.5	9.4	8.7
2PSZB	8.4	8.0	5.2	5.8	5.4	6.9

Table 3.29 Peak side forces in the third phase of movement of the platform through the bottom level of the man-material cage in the six experimental cycles

Marking of the Force	*Force Value (kN)*					
	1. Cycle	*2. Cycle*	*3. Cycle*	*4. Cycle*	*5. Cycle*	*6. Cycle*
3PNWB	12.4	11.9	11.4	11.2	13.7	13.3
3PNZB	10.8	10.4	9.0	11.1	9.5	9.9
3PSWB	7.2	9.1	6.8	5.3	8.2	8.5
3PSZB	7.3	11.2	4.7	6.9	9.1	5.8

Table 3.30 Peak side forces in the fourth phase of movement of the platform through the bottom level of the man-material cage in the six experimental cycles

Marking of the Force	*Force Value (kN)*					
	1. Cycle	*2. Cycle*	*3. Cycle*	*4. Cycle*	*5. Cycle*	*6. Cycle*
4PNWB	13.9	16.4	15.8	16.7	14.4	17.2
4PNZB	11.5	15.0	10.3	12.2	9.3	13.9
4PSWB	14.9	16.2	17.1	16.0	15.5	14.3
4PSZB	17.8	16.5	13.2	15.9	17.2	14.8

Figures 3.24 and 3.25 present graphs of the peak forces' distribution for each phase of the test:

- graphs from Figure 3.24 present peak forces caused by the man-material cage and acting on the stiff guides on the northern side,
- graphs shown in Figure 3.25 present peak forces caused by the conveyance and acting on the stiff guides on the southern side.

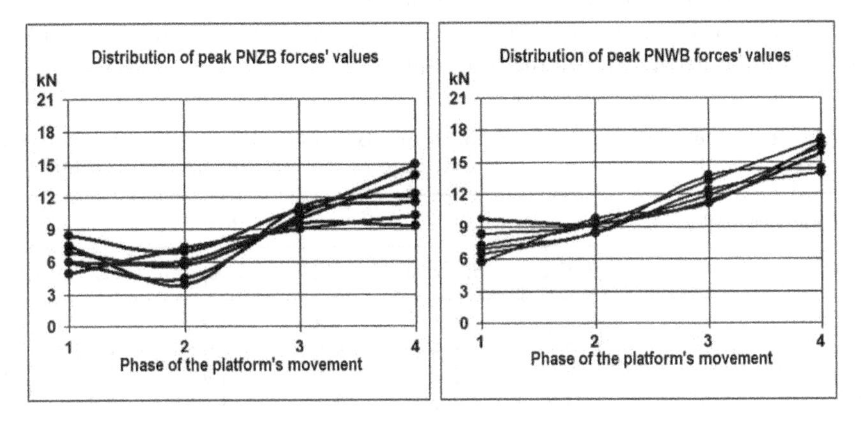

Figure 3.24 Diagrams of distributions of peak side forces' values acting on the northern bottom retractable guides on the level 960 m in different phases of the platform's movement through the bottom level of the cage.

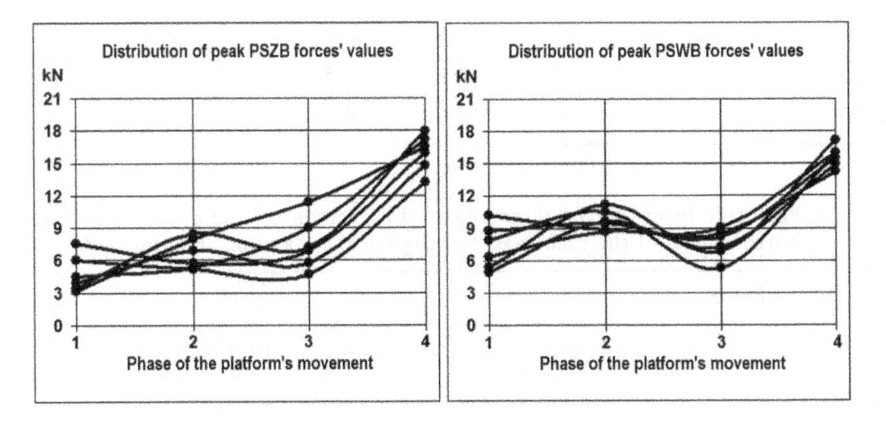

Figure 3.25 Diagrams of distributions of peak side forces' values acting on the southern bottom retractable guides on the level 960 m in different phases of the platform's movement through the bottom level of the cage.

3.2.8 Top transom side forces

Peak side forces of the top transom, labelled with markings shown in Figure 3.5 are presented in Tables 3.31–3.34, of which Table 3.31 presents forces measured in the first phase of test, Table 3.32 – in the phase 2, Table 3.33 – in the phase 3 and Table 3.34 – in the phase 4.

Figures 3.26 and 3.27 present graphs of the peak forces' distribution for each phase of the test:

- graphs from Figure 3.26 present peak forces caused by the man-material cage and acting on the stiff guides on the northern side,
- graphs shown in Figure 3.27 present peak forces caused by the conveyance and acting on the stiff guides on the southern side.

Table 3.31 Peak side forces in the first phase of movement of the platform through the bottom level of the man-material cage in the six experimental cycles

Marking of the Force	*Force Value (kN)*					
	7. Cycle	*8. Cycle*	*9. Cycle*	*10. Cycle*	*11. Cycle*	*12. Cycle*
1GNWB	3.2	5.1	3.8	3.7	4.9	3.3
1GNZB	7.1	6.9	5.5	4.9	4.1	4.5
1GSWB	3.1	3.8	2.5	3.9	3.2	2.3
1GSZB	3.9	2.2	4.4	3.1	4.4	5.2

Table 3.32 Peak side forces in the second phase of movement of the platform through the bottom level of the man-material cage in the six experimental cycles

Marking of the Force	*Force Value (kN)*					
	7. Cycle	*8. Cycle*	*9. Cycle*	*10. Cycle*	*11. Cycle*	*12. Cycle*
2GNWB	4.5	5.6	3.7	3.2	4.3	3.9
2GNZB	5.5	3.7	5.2	4.7	5.0	4.6
2GSWB	4.9	6.5	3.2	5.8	7.5	4.2
2GSZB	6.1	7.4	4.7	5.5	5.1	4.2

Table 3.33 Peak side forces in the third phase of movement of the platform through the bottom level of the man-material cage in the six experimental cycles

Marking of the Force	*Force Value (kN)*					
	7. Cycle	*8. Cycle*	*9. Cycle*	*10. Cycle*	*11. Cycle*	*12. Cycle*
3GNWB	6.9	8.1	7.8	7.5	7.6	9.3
3GNZB	7.5	5.1	5.3	5.9	6.2	4.7
3GSWB	5.5	5.1	3.2	3.9	4.7	2.9
3GSZB	5.1	5.9	4.8	3.5	7.1	2.9

Table 3.34 Peak side forces in the fourth phase of movement of the platform through the bottom level of the man-material cage in the six experimental cycles

Marking of the Force	*Force Value (kN)*					
	7. Cycle	*8. Cycle*	*9. Cycle*	*10. Cycle*	*11. Cycle*	*12. Cycle*
4GNWB	10.4	9.1	8.7	11.1	9.3	10.5
4GNZB	11.9	10.7	11.2	10.6	12.3	10.8
4GSWB	13.2	12.7	11.9	14.1	15.2	14.3
4GSZB	12.9	12.5	11.7	14.1	12.3	11.0

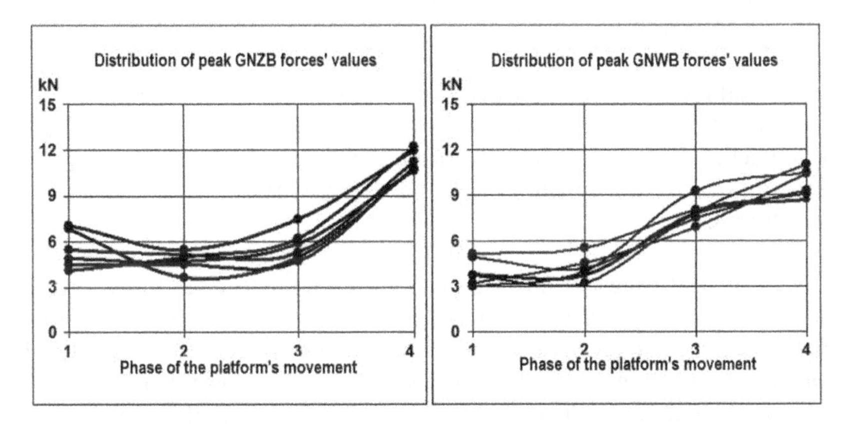

Figure 3.26 Diagrams of distributions of peak side forces' values acting on the northern top retractable guides on the level 960 m in different phases of the platform's movement through the bottom level of the cage.

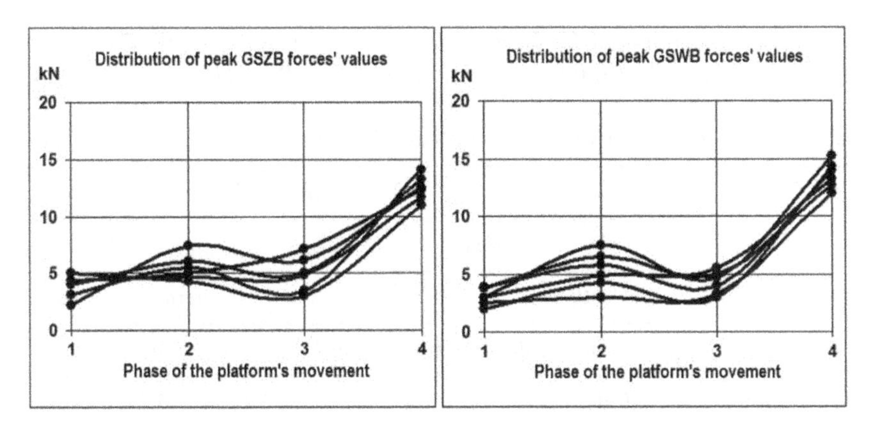

Figure 3.27 Diagrams of distributions of peak side forces' values acting on the southern top retractable guides on the level 960 m in different phases of the platform's movement through the bottom level of the cage.

3.2.9 Physical verification of measured forces

Physical verification utilizes the method presented in ISO (1995) and recommended by Polish Central Office of Measures. This method consists of the so-called type B uncertainty of measure. It is calculated using the following formula:

$$\lambda = \frac{|L_0 - U|}{L_0 + U} * 100\% \tag{3.1}$$

where

L_0, mm – clearance between retractable guide front and rubbing block on the conveyance, $L_0 = 5\,mm$;

U, mm – displacement of the rubbing block against the retractable guide front, corresponding to the peak force measured.

Table 3.35 consists of the following:

- peak frontal forces for each retractable guide from Tables 3.19–3.22,
- displacements of rubbing blocks against the retractable guides, calculated according to Płachno (2005) by processing of measuring signals,
- uncertainty of measurement, calculated using formula 3.1 for each of the presented forces.

Table 3.35 shows that values of uncertainty λ for each of the peak forces measured for retractable guides do not exceed 15%, which means that the acceptable level of the measurement error was not exceeded. Thus, measured forces can be considered a reliable projection of real forces acting on the construction of the retractable guidance system caused by loading and unloading of the man-material cage with the maximum allowed load.

3.2.10 Stress in the bridle hangers of the man-material cage caused by the bottom section of the retractable guidance system

Peak values of stress in bridle hangers of the man-material cage caused by the bottom section of the retractable guidance system are presented in Tables 3.36–3.39, of which

Table 3.35 Parameters of verification of measured forces

Parameter	*Retractable Guide*							
	Entry				*Exit*			
	Northern		*Southern*		*Northern*		*Southern*	
	Bottom	*Top*	*Bottom*	*Top*	*Bottom*	*Top*	*Bottom*	*Top*
Peak frontal force, kN	40.8	34.9	38.3	36.3	38.9	39.7	40.7	40.2
Displacement U, mm	6.2	4.1	5.5	4.9	5.6	5.9	6.1	6.0
Uncertainty λ, %	10.7	9.9	4.8	8.7	5.7	8.2	9.9	9.1

Table 3.36 Peak forces of stress in man-material cage's bridle hangers caused by forces occurring in the first phase of movement of the platform through the bottom level of the cage in the six experimental cycles

Marking of the Force Causing Stress	*Stress Value (kN)*					
	1. Cycle	*2. Cycle*	*3. Cycle*	*4. Cycle*	*5. Cycle*	*6. Cycle*
1PNW	27.6	49.0	23.8	59.8	33.3	28.1
1PNZ	28.7	61.9	37.3	51.7	51.1	49.6
1PSW	30.7	31.8	38.1	42.1	33.9	24.6
1PSZ	39.0	32.8	22.4	23.1	16.7	15.1

Table 3.37 Peak forces of stress in man-material cage's bridle hangers caused by forces occurring in the second phase of movement of the platform through the bottom level of the cage in the six experimental cycles

Marking of the Force Causing Stress	*Stress Value (kN)*					
	1. Cycle	*2. Cycle*	*3. Cycle*	*4. Cycle*	*5. Cycle*	*6. Cycle*
2PNW	62.1	46.6	56.3	50.3	59.7	53.9
2PNZ	47.3	30.9	44.8	40.0	26.3	56.3
2PSW	42.9	55.9	59.6	53.4	44.2	51.4
2PSZ	33.4	36.3	34.0	39.7	44.0	48.8

Table 3.38 Peak forces of stress in man-material cage's bridle hangers caused by forces occurring in the third phase of movement of the platform through the bottom level of the cage in the six experimental cycles

Marking of the Force Causing Stress	*Stress Value (kN)*					
	1. Cycle	*2. Cycle*	*3. Cycle*	*4. Cycle*	*5. Cycle*	*6. Cycle*
3PNW	79.4	67.6	69.4	64.6	83.2	59.3
3PNZ	70.4	66.2	80.4	82.6	73.8	78.2
3PSW	53.8	40.8	30.9	59.2	33.8	61.4
3PSZ	37.3	65.8	18.6	15.2	36.7	47.1

Table 3.39 Peak forces of stress in man-material cage's bridle hangers caused by forces occurring in the fourth phase of movement of the platform through the bottom level of the cage in the six experimental cycles

Marking of the Force Causing Stress	*Stress Value (kN)*					
	1. Cycle	*2. Cycle*	*3. Cycle*	*4. Cycle*	*5. Cycle*	*6. Cycle*
4PNW	105.6	94.2	97.3	100.3	86.8	84.4
4PNZ	97.8	103.7	106.5	84.8	99.2	90.7
4PSW	95.9	87.1	109.1	102.4	93.9	90.1
4PSZ	108.5	98.2	102.9	92.5	100.3	94.2

Table 3.36 presents stress measured in the first phase of test, Table 3.37 – in the phase 2, Table 3.38 – in the phase 3 and Table 3.39 – in the phase 4.

Figures 3.28 and 3.29 show graphs of peak stress value distribution resulting from the movement of the platform through the bottom level of the man-material cage:

- graphs in Figure 3.28 present stress in the northern bridle hangers of the conveyance, both on the entry and exit,
- graphs in Figure 3.29 present stress in the southern bridle hangers of the conveyance, both on the entry and exit.

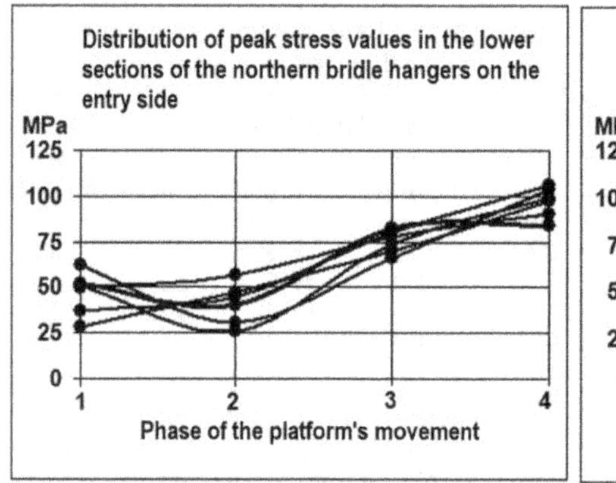

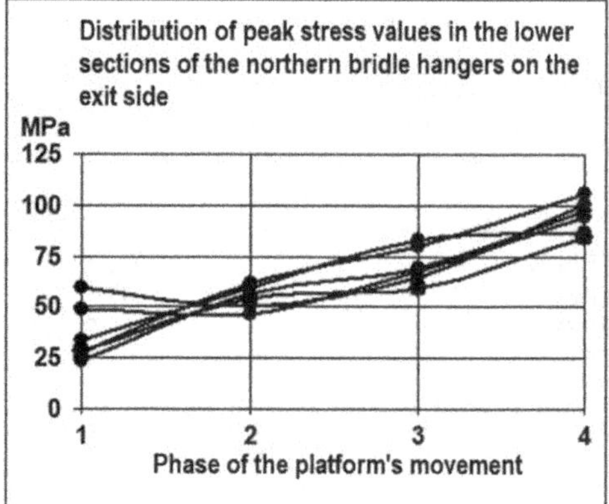

Figure 3.28 Diagrams of distributions of peak stress values in the lower sections of the northern bridle hangers of the man-material cage.

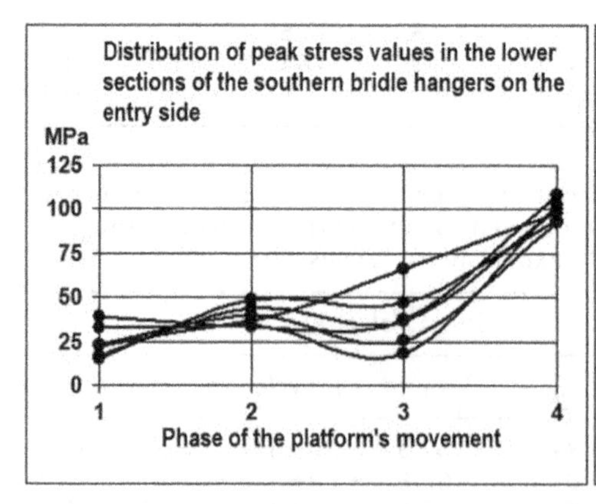

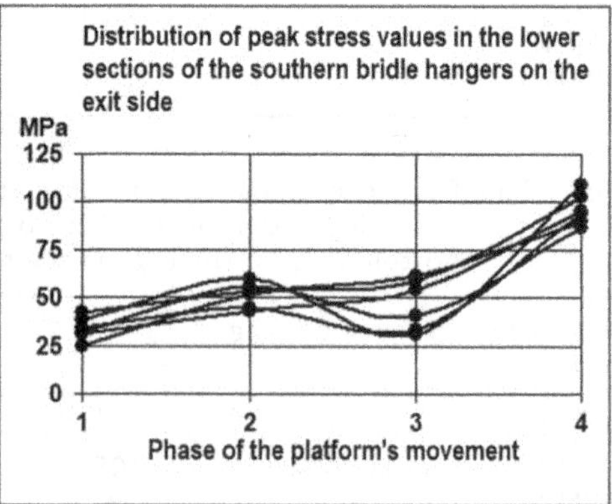

Figure 3.29 Diagrams of distributions of peak stress values in the lower sections of the southern bridle hangers of the man-material cage.

3.2.11 Stress in the bridle hangers of the man-material cage caused by the top section of the retractable guidance system

Peak values of stress in bridle hangers of the man-material cage caused by the top section of the retractable guidance system are presented in Tables 3.40–3.43, of which Table 3.40 presents stress measured in the first phase of test, Table 3.41 – in the phase 2, Table 3.42 – in the phase 3 and Table 3.43 – in the phase 4.

Figures 3.30 and 3.31 show graphs of peak stress value distribution resulting from the movement of the platform through the bottom level of the man-material cage:

- graphs in Figure 3.30 present stress in the northern bridle hangers of the conveyance, both on the entry and exit,

Table 3.40 Peak forces of stress in man-material cage's bridle hangers caused by forces occurring in the first phase of movement of the platform through the bottom level of the cage in the six experimental cycles

Marking of the Force Causing Stress	*Stress Value (kN)*					
	7. Cycle	*8. Cycle*	*9. Cycle*	*10. Cycle*	*11. Cycle*	*12. Cycle*
1GNW	47.9	24.0	20.4	19.7	50.6	29.9
1GNZ	43.9	26.7	20.5	46.1	29.5	44.3
1GSW	42.1	24.1	43.2	23.8	25.6	25.3
1GSZ	24.7	17.9	38.1	20.3	34.1	27.2

Table 3.41 Peak forces of stress in man-material cage's bridle hangers caused by forces occurring in the second phase of movement of the platform through the bottom level of the cage in the six experimental cycles

Marking of the Force Causing Stress	*Stress Value (kN)*					
	7. Cycle	*8. Cycle*	*9. Cycle*	*10. Cycle*	*11. Cycle*	*12. Cycle*
2GNW	36.3	27.7	28.3	41.4	55.8	35.2
2GNZ	17.2	26.9	34.3	30.7	38.3	34.2
2GSW	62.8	46.5	57.2	53.0	81.5	36.3
2GSZ	70.7	44.4	49.3	49.2	28.3	39.5

Table 3.42 Peak forces of stress in man-material cage's bridle hangers caused by forces occurring in the third phase of movement of the platform through the bottom level of the cage in the six experimental cycles

Marking of the Force Causing Stress	*Stress Value (kN)*					
	7. Cycle	*8. Cycle*	*9. Cycle*	*10. Cycle*	*11. Cycle*	*12. Cycle*
3GNW	83.6	48.6	68.9	44.8	62.1	47.1
3GNZ	40.2	51.4	70.3	41.8	39.2	45.8
3GSW	52.4	49.7	58.9	53.2	58.3	42.9
3GSZ	50.8	40.3	11.3	58.3	18.5	44.8

Table 3.43 Peak forces of stress in man-material cage's bridle hangers caused by forces occurring in the fourth phase of movement of the platform through the bottom level of the cage in the six experimental cycles

Marking of the Force Causing Stress	*Stress Value (kN)*					
	7. Cycle	*8. Cycle*	*9. Cycle*	*10. Cycle*	*11. Cycle*	*12. Cycle*
4GNW	117.1	106.3	98.9	129.8	121.9	93.8
4GNZ	100.2	97.6	118.2	93.1	108.3	123.1
4GSW	120.8	102.2	94.8	105.3	128.3	114.2
4GSZ	104.3	119.1	97.3	109.8	116.6	126.9

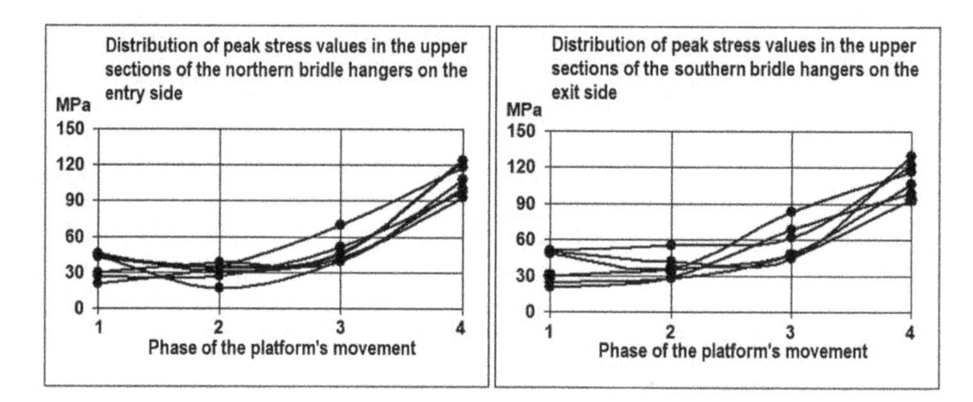

Figure 3.30 Diagrams of distributions of peak stress values in the upper sections of the northern bridle hangers of the man-material cage.

- graphs in Figure 3.31 present stress in the southern bridle hangers of the conveyance, both on the entry and exit.

3.2.12 Physical verification of measured stress

Similar to the previously presented forces' verification, the method described in ISO (1995) and recommended by Polish Central Office of Measures was used. For the issue of measured stress, the following formula is used to calculate the uncertainty of measurement:

$$\lambda = \frac{|\sigma_Z - \sigma_O|}{\sigma_Z + \sigma_O} * 100\% \tag{3.2}$$

where

σ_Z, MPa – peak measured stress,

σ_O, MPa – stress calculated according to the calculation model presented in Płachno (2018a), corresponding to measured forces and displacements of the floor beam of the second cage level.

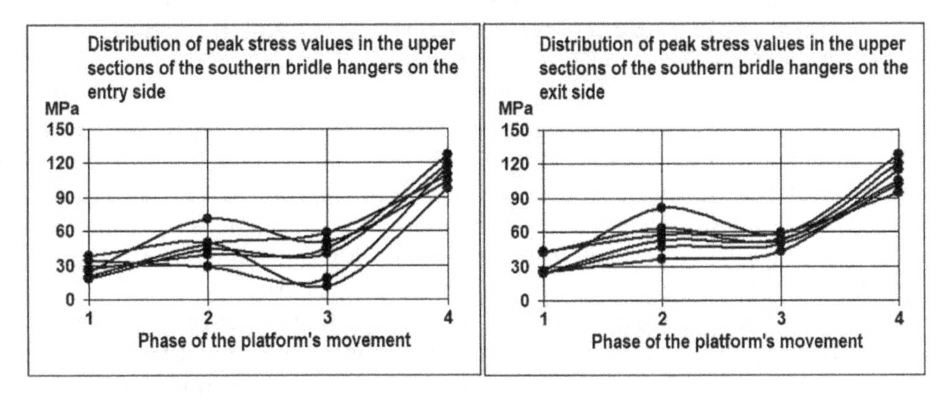

Figure 3.31 Diagrams of distributions of peak stress values in the upper sections of the southern bridle hangers of the man-material cage.

Table 3.44 Parameters of verification of measured stress

Parameter	*Part of the Conveyance's Bridle Hanger*							
	Entry				*Exit*			
	Northern		*Southern*		*Northern*		*Southern*	
	Bottom	*Top*	*Bottom*	*Top*	*Bottom*	*Top*	*Bottom*	*Top*
Peak measured stress σ_z, MPa	106.5	123.8	108.5	126.9	105.6	129.8	109.1	128.3
Design stress σ_o, MPa	133.9	104.0	134.8	107.9	129.6	108.6	135.0	107.5
Uncertainty λ, %	11.4	8.7	10.8	8.1	10.2	8.9	10.6	8.8

Table 3.44 presents the following:

- peak verified stress σ_z from Tables 3.36–3.39 for each section of bridle hangers of the man-material cage,
- calculated stress σ_0, corresponding to the measured forces and displacements of the floor beam of the cage's upper level,
- uncertainty of measurement λ, calculated according to the formula (3.2) for verified stress σ_z.

Table 3.44 shows that uncertainty of measurement λ of the biggest values of stress, measured in 12 experimental cycles of loading and unloading the conveyance, in cage's bridle hangers caused by the construction of the retractable guidance system does not exceed 15%, which means that the measurement error is on an acceptable level. Thus, measured stress can be considered a reliable projection of real stress occurring in bridle hangers of the conveyance, caused by loading and unloading of the man-material cage with the maximum allowed load.

3.2.13 Comparison of forces measured for stiff guides and for the retractable guidance system on the level 960 m in the Leon IV shaft

Comparative analysis of forces measured in the tests of stiff guidance and retractable guidance used for the man-material cage on the level 960 m in the Leon IV shaft. For the purpose of analysis, peak values of compared forces are listed in Tables 3.45 and 3.46. Peak forces obtained in measurements of retractable guidance system match data presented in Tables 3.21–3.35. and forces measured for the stiff guidance are shown in Tables 3.17 and 3.18 (Kamiński, Prostański and Dyczko 2021).

Table 3.45 shows that peak frontal forces caused by the bottom transom of the man-material cage and acting on the retractable guides were equal to 40.8 kN and on the stiff guides 48.3 kN, respectively, and forces caused by the top transom for the retractable guides were equal 40.2 kN and for the stiff guides 48.2 kN.

Respectively, Table 3.46 shows that peak side forces caused by the bottom transom and acting on the retractable guides were equal to 17.9 kN, whereas side forces acting on the stiff guidance were equal to 20.1 kN. Top transom caused forces equal to 15.2 kN, acting on the retractable guides and 18.4 on the stiff guides.

Table 3.45 Peak frontal forces acting on the guides on the level 960 m in the Leon IV shaft

Marking of the Force	*Peak Frontal Forces Caused by the Bottom Transom of the Man-Material Cage Acting on the Guides*		*Peak Frontal Forces Caused by the Top Transom of the Man-Material Cage Acting on the Guides*	
	Retractable Guides (kN)	*Stiff Guides (kN)*	*Retractable Guides (kN)*	*Stiff Guides (kN)*
NWC	38.9	48.3	39.7	44.3
NZC	40.8	35.9	34.9	31.4
SWC	40.7	36.2	40.2	48.2
SZC	38.3	40.1	36.3	36.7

Table 3.46 Peak side forces acting on the guides on the level 960 m in the Leon IV shaft

Marking of the Force	*Peak Side Forces Caused by the Bottom Transom of the Man-Material Cage Acting on the Guides*		*Peak Side Forces Caused by the Top Transom of the Man-Material Cage Acting on the Guides*	
	Retractable Guides (kN)	*Stiff Guides (kN)*	*Retractable Guides (kN)*	*Stiff Guides (kN)*
NWB	17.2	20.1	11.0	11.5
NZB	15.0	15.4	12.2	12.5
SWB	17.1	16.7	15.2	18.4
SZB	17.9	19.8	14.1	9.5

Based on the presented analysis of the measured forces, one can conclude that the values of both peak frontal and side forces caused by bottom and top transom of the man-material cage acting on the construction of the retractable guidance system are significantly lower than forces acting on the stiff guidance used on the level 960 m in the Leon IV shaft. Such reduction of obtained forces' values may be a result of lower value of the clearance between the retractable guide and the conveyance that between the cage and the stiff guides (Kamiński, Prostański and Dyczko 2021; PN-M-06515).

3.2.14 Comparison of forces measured for the retractable guidance system and existing legislation in terms of safety and Polish Standard PN-G-46227. Mine shafts: Chairing: Requirements

3.2.14.1 Design forces for the retractable guidance system according to the standards

Polish standard PN-G-46227. Mine shafts: Chairing: Requirements (PN-G-46227) does not include definition of the term "retractable guidance system", but analogy between functions of chairing system (defined in the Standard) and retractable guidance system and requirements for chairing system should be considered during measurements and evaluation of the retractable guidance system construction.

The Standard (PN-G-46227) consists of the following requirements for design forces acting on shaft chairing construction (2.1.1.3):

> Horizontal load acting on shaft chairing construction during loading a cage with mine cart should be taken equal to peak force acting on tub creeper's driver, but not less than 10 kN.
>
> Load-bearing elements of shaft chairing should be characterized with at least six times higher safety factor, without taking into account wear limit.

Cited paragraphs should be considered sufficient for the purpose of determination of the design forces of the shaft chairing construction, but only if the system has got a typical form of construction. However, construction of the retractable guidance system can be considered a shaft chairing system only in terms of its functions; thus, determination of design forces according to requirements of the Standard (PN-G-46227) involves additional assumptions for both tub creeper driver's load distribution on different elements of retractable guidance system and selection of design side force acting on these elements.

An attempt to prove the existence of the mentioned load distribution was made by comparing peak frontal forces acting on retractable guides and caused by tub creeper – determined as 20 kN (Zaprojektowanie 2005) with measured peak forces (Table 3.27). However, this load distribution's existence was not clearly proven. It suggests that peak force caused by tub creeper should be assumed equal to peak design force acting on each of eight retractable guides installed on the level 960 m in the Leon IV shaft.

In terms of side forces acting on the retractable guidance system's elements, it should be assumed that their construction, different than typical shaft chairing's construction, precludes a priori assumption of compliance of the strength requirements of the side forces, based on previously presented compliance in case of the frontal forces. Analysis of side forces revealed that the greatest value of side force acting on the retractable guides occurs in the case of derailment of the transport unit, which had not happened during tests. However, peak side force value in case of the derailment can be estimated based on the measurement's results. According to this estimation, peak side force value is about 70% higher than the measured peak side force, and peak frontal force is about 30% lower than measured peak frontal force (Kamiński, Prostański and Dyczko 2021).

According to facts and assumptions presented above, it was considered reasonable to include two cases during calculation of the forces acting on the retractable guidance system (Kamiński, Prostański and Dyczko 2021):

- case 1, consisting of design force calculated according to the Standard (PN-G-46227), as the peak force acting on the tub creeper's driver. However, calculations are conducted for the case of this force acting on each of the retractable guides,
- case 2, comprising the transport unit's derailments, with simultaneously acting peak frontal and side forces, equal to, respectively, 70% of peak force on the tub creeper's driver, acting on each of the retractable guides.

3.2.15 Evaluation of measured forces acting on the retractable guidance system

The base of the evaluation presented below are the safety factor specified in the Standard (PN-G-46227) for the shaft chairing's construction and in Appendix 4 to Regulation (rozporządzenie 2016), in which shaft chairing is considered "auxiliary equipment of the mine shaft".

The required safety factor, presented both in the Standard (PN-G-46227) and Regulation (rozporządzenie 2016), is characterized with equal value, but Section 3.15 of Appendix 4 to the Regulation (rozporządzenie 2016) makes this factor a criterium for design stress verification, specifying the following:

- definition of load-bearing elements' safety factor calculation, as the ratio of the tensile strength R_m of the material and the design stress,
- legitimacy of the Standard's (PN-G-46227) record, which states that the wear limit of shaft chairing's elements is not included in their design stresses,
- requirement for corrosion and wear allowance for steel elements of shaft chairing's construction.

Based on the Standard (PN-G-46227) and requirements of the Regulation (rozporządzenie 2016), the obligatory safety factor for load-bearing elements of the shaft chairing was defined as a product of three coefficients:

- n_d – coefficient of frontal dynamic loads acting on the shaft chairing construction.
- n_g – coefficient of limit stress of load-bearing elements of the shaft chairing,
- n_z – coefficient of wear limit of shaft chairing elements.

Coefficient n_d is defined as the ratio of peak frontal force acting on the retractable guides (Table 3.45) and peak force measured on the tub creeper's driver (Zaprojektowanie 2005), which is the cause of the force acting on the guides. The calculated value of the n_d coefficient is equal to 2.0.

Coefficient n_g was defined because the limit stress in the retractable guides is equal to the yield point of the material of which they are made, but verification of stress in the guides should be done based on its tensile strength, according to the Regulation (rozporządzenie 2016). Thus, the value of coefficient n_g was accepted at the level of 1.8.

Finally, the coefficient n_z is calculated as the ratio of required safety factor and product of coefficients n_d and n_g. According to this, the value of the coefficient n_z is equal to 1.67.

Based on the values of the coefficients presented above, the following note of the forces measured on the retractable guidance system was formulated:

1. Peak frontal forces acting on the retractable guidance system, treated as shaft chairing (based on the existing Regulation (rozporządzenie 2016) and Standard (PN-G-46227)), measured in tests can be considered safe (according to the Regulation [rozporządzenie 2016]), if each of the load-bearing elements of its construction is characterized with safety factor not less than 6 for the peak force caused by the tub creeper's driver acting on each of the eight retractable guides installed on the level 960 m in the Leon IV shaft.
2. Peak side forces acting on the retractable guidance system, treated as shaft chairing, measured in tests can be considered safe (according to the Regulation [rozporządzenie 2016]), if each of the load-bearing elements of its construction is characterized with safety factor not less than six for the simultaneously acting peak frontal and side forces, equal to, respectively, 70% of peak force on the tub creeper's driver, acting on each of the retractable guides.
3. Based on the calculated safety factors presented above, corrosion and wear allowance (according to Section 3.15.1 of the Appendix 4. to the Regulation [rozporządzenie 2016]) can be determined for sections used as the load-bearing elements of the retractable guidance system's construction. To calculate this allowance, the following formula is recommended to be used:

$$\Delta = 1.83 * n_{min}^2 - 5.14 * n_{min} - 5.18 \tag{3.3}$$

where
Δ – wear limit, as a part of initial section wall thickness, %,
n_{min} – the smallest safety factor calculated for load-bearing elements of the retractable guidance system's construction.

3.2.16 Evaluation of stress in the bridle hangers of the man-material cage caused by the retractable guides

Calculations presented in the following section were made to verify the assumption that the application of the retractable guidance system allows resign of the stabilization in plane of the floor of the upper level of the man-material cage, previously used during operation of stiff guidance on the level 960m in the Leon IV shaft. This assumption was supported by a statement that fatigue limit of the conveyance's bridle hangers, corresponding to measured stress, is at least equal to the amount of loads and unloads of the cage with the transport unit of the conveyance's maximum load within the whole cage's lifetime.

The number of loads and unloads of the man-material cage was estimated at the level of 150 thousand, because of the conveyance's lifetime equal to 20 years and average number of loads equal ten per day and identical amount of unloads, in both cases with the conveyance's maximum load.

For the purpose of the verification of the fulfilment of the endurance requirement of the cage's bridle hangers, the calculation method presented in the Standard (PN-M-06515) was applied. This method is used in practice in determining the durability of bridle hangers used in skips (Płachno 2018a).

The Standard (PN-M-06515) states that needed endurance of the cage's bridle hangers is ensured if the peak stress caused by the retractable guides in each of the eight parts of the bridle hangers presented in the Table 3.47 fulfils the following condition:

$$\sigma_{\max_i} \leq R_{zi} = 205 - 87 * K_{pi}, \quad i = 1,2,\ldots,8 \tag{3.4}$$

where

R_{zi} – fatigue strength of the i-th part of the cage's bridle hanger, equal expected endurance of bridle hangers, corresponding to measured stress, caused by the retractable guides, according to the standard (PN-M-06515) MPa,

K_{pi} – the so-called load parameter (according to the Standard PN-M-06515), calculated using the formula (15) introduced in this Standard.

For the purpose of calculating stress in the current case, formula (15) presented in the Standard (PN-M-06515) applies as follows:

$$K_{pi} = \sqrt[7]{\sum_{j=1}^{j=N_c} \left(\frac{\sigma_{ij}}{\sigma_{\max ij}} \right)^7 * \frac{1}{N_c}}, \quad j = 1,2,3,\ldots,N_c \tag{3.5}$$

where

N_c – number of experimental cycles of the platform movement through the bottom level of the man-material cage in the shaft during the measurements of stress, equal to 12,

σ_{ij} – measured stress (matching data presented in Tables 3.36–3.43), which is equal to the measured value if it fulfils the following condition:

$$\sigma_{ij} \geq R_w \tag{3.6}$$

and is equal to zero in other cases.

Presented in the formula (3.6) parameter R_w, called in the Standard (PN-M-06515) permanent endurance of the load-bearing element, according to the Standard, can be calculated for each of the conveyance's bridle hangers, if their connections with the cage's elements are assigned using a proper type of the notch. Based on the above, the value of the endurance R_w is equal to 84 MPa.

Table 3.47 Parameters of fatigue limit of bridle hangers' parts

Parameter of Fatigue Limit	*Entry*				*Exit*			
	Northern		*Southern*		*Northern*		*Southern*	
	Bottom	*Top*	*Bottom*	*Top*	*Bottom*	*Top*	*Bottom*	*Top*
K_{pi}	0.84	0.81	0.84	0.82	0.83	0.81	0.82	0.81
R_{zi}, MPa	131.9	134.5	131.9	133.7	132.8	134.5	133.7	134.5
σ_{maxi}, MPa	106.5	123.8	108.5	126.9	105.6	129.8	109.1	128.3
σ_{maxi}/R_{zi}	0.81	0.92	0.82	0.95	0.80	0.96	0.82	0.95

Table 3.47 presents the values of the K_{pi} coefficient, calculated based on the formulas (3.5) and (3.6) as well as the following:

- Value of endurance limit R_{zi} for each of the eight parts of the conveyance's bridle hangers, calculated according to the formula (3.4) for each of the K_{pi} coefficients,
- stress σ_{maxi}, matching the data presented in the Tables 3.36–3.43,
- ratio of stress σ_{maxi} and endurance limit R_{zi}.

Based on the endurance parameters presented in Table 3.47 of each of the part of the man-material cage's bridle hangers, the following note of stress measured in these bridle hangers and caused by the retractable guides was formulated:

1. Each of the eight parts of the man-material cage's bridle hangers in the Leon IV shaft is characterized by the ratio of measured stress and fatigue endurance less than 1. It follows that the condition of endurance of the conveyance's bridle hangers is met.
2. The assumption that resignation of the additional stabilization of the man-material cage in the plane of the floor of its upper level was empirically proven right. It follows that possible fatigue damage of the conveyance's bridle hangers during its operation in the Leon IV shaft is an effect of the stress rise in them or other factors, not related with the construction of the retractable guidance system.

3.2.17 Final conclusions of the retractable guidance system measurements

1. Physical verification of quantities measured in presented tests, which are peak frontal forces acting on the retractable guides and peak stress occurring in the bridle hangers of the man-material cage, revealed that both measured forces and stresses are a satisfactorily reliable image of real forces and stresses occurring in the elements of the retractable guidance system and the cage in the process of loading and unloading of the cage on the level 960 m in the Leon IV shaft with the transport units of the maximum load possible for the conveyance.

2. Presented tests demonstrate that
   - peak frontal forces acting on the elements of the retractable guidance system, treated as auxiliary shaft equipment (rozporządzenie 2016), can be considered safe, according to the existing law, if value of the design safety factor is not less than six, for the maximum possible force caused by the tub creeper's driver acting on each of the load-bearing elements of the retractable guidance system on the level 960 m in the Leon IV shaft,
   - peak side forces acting on the retractable guidance system, treated as shaft chairing, measured in tests can be considered safe (according to the Regulation [rozporządzenie 2016]), if each of the load-bearing elements of its construction is characterized with safety factor not less than six for the simultaneously acting peak frontal and side forces, equal to, respectively, 70% of peak force on the tub creeper's driver, acting on each of the retractable guides.
   - Based on the calculated safety factors presented above, corrosion and wear allowance (according to Section 3.15.1 of Appendix 4 to the Regulation [rozporządzenie 2016]) can be determined for sections used as the load-bearing elements of the retractable guidance system's construction. To calculate this allowance, introduced in the Standard (PN-M-06515) and presented in this work, formula (3.3) is recommended to be used.
3. Analysis of the fatigue strength of the bridle hangers of the man-material's cage in the Leon IV shaft, including an influence of the stress measured in bridle hangers and caused by the retractable guides, presented above, demonstrate the following:
   - Each of the eight parts of the man-material cage's bridle hangers in the Leon IV shaft is characterized by the ratio of measured stress and fatigue endurance less than 1. It follows that the condition of endurance of the conveyance's bridle hangers is met, even though no additional support of the conveyance's elements is used,
   - the assumption that resignation of the additional stabilization of the man-material cage in the plane of the floor of its upper level was empirically proven right. It follows that possible fatigue damage of the conveyance's bridle hangers during its operation in the Leon IV shaft is an effect of the stress rise in them or other factors, not related with the construction of the retractable guidance system.

3.3 MECHANICAL ANALYSIS OF CONSTRUCTION OF THE RETRACTABLE GUIDANCE SYSTEM

Mechanical analysis of the retractable guidance system applied on the level 960 m in the Leon IV shaft covers the following scope:

1. Computational schemes and calculation of forces acting on the construction and particular elements of the retractable guidance system during its movement from resting to working position.
2. Computational schemes and calculation of forces acting on the construction and particular elements of the retractable guidance system during its work.

3.3.1 Diagrams and calculations of forces acting on the bottom section of the retractable guidance system during its movement from resting to working position

Diagrams of forces acting on the bottom section of the retractable guidance system are shown in Figure 3.32. Subsequent diagrams, marked as (a–d), present forces acting on the construction of the retractable guidance system in the following moments of its movement from resting to working position:

a. the beginning of the section's movement; zero moment, which is the beginning of the bottom section's wheels' movement along the inclined race,
b. moment '1', which is the finish of bottom section's wheels' movement along the inclined race,
c. moment '2', which is the beginning of bottom section's wheels' movement along the vertical race,
d. moment '1', which is the finish of bottom section's wheels' movement along the vertical race; the end of the movement to the working position.

The following forces acting on the bottom section of the retractable guidance system are presented in Figure 3.32:

- **G** – weight of the bottom section of the retractable guidance system,
- **Q0, Q2, Q3, Q4** – force acting on the axis of the bottom wheels of the bottom section of the retractable guidance system, respectively, in moments 0, '1', '2' and '3',

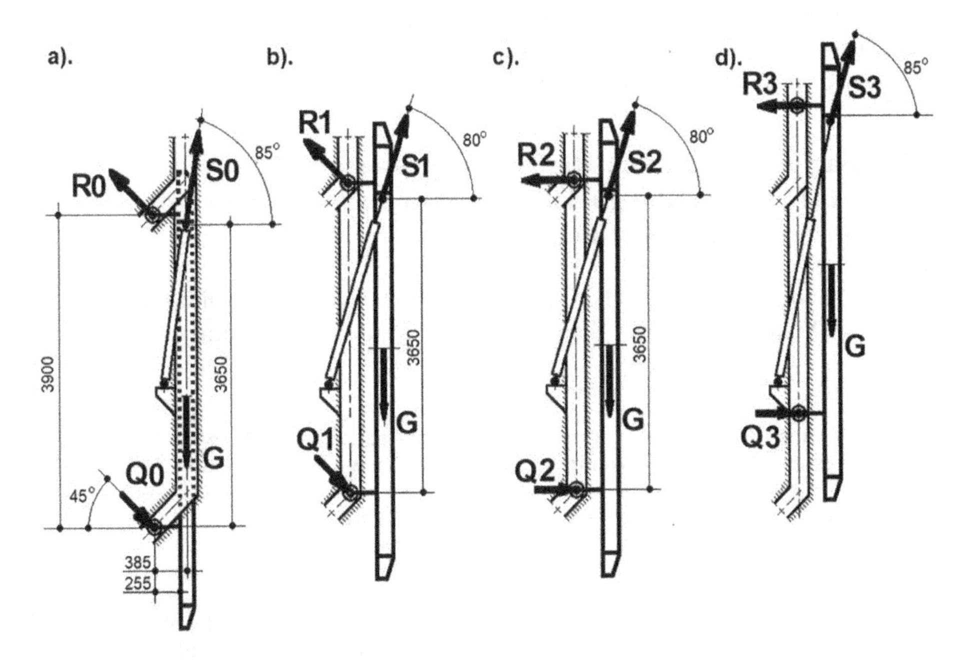

Figure 3.32 Diagrams of the forces acting on the bottom section of the retractable guidance system.

- **R0, R1, R2, R3** – force acting on the axis of the top wheels of the bottom section of the retractable guidance system, accordingly in moments 0, '1', '2' and '3',
- **S0, S1, S2, S3** – force caused by the drive of the bottom section of the retractable guidance system, respectively, in moments 0, '1', '2' and '3'.

Diagrams presented in Figure 3.32 were drawn up on the basis of Projekt techniczny (2018). Dimensions presented in Figure 3.32 were referred to the mentioned project, and the value of the weight G was calculated as equal to 19 kN.

Table 3.48 presents the results of calculations of the forces Q, R and S values. The assumption was made that both inertial forces and friction acting on the bottom section of the retractable guidance system during its movement from resting to working position are negligibly small.

Table 3.48 shows that values of the forces Q, R and S in the bottom section of the retractable guidance system during its motion from resting to working position are significantly different from each other. The smallest force is Q, with the peak value about 1.0 kN occurring during wheels' movement upside the inclined race, while in the moment of wheels' entry on the vertical race, the value of the force Q drops immediately to its minimum, which is about 0.4 kN. Then, during its motion upside the vertical race, it constantly raises to the value 0.7 kN at the end of the race.

In turn, force S is characterized by the greatest value, which in the initial moment is equal 18.4 kN, and during wheels' motion along the inclined race, it is decreasing to 18.3 kN. In the moment of wheels' entry on the vertical race, the value of the force S reaches 20.3 kN, and during their movement upside the race, force S decreases to 20.1 kN in the final moment.

The initial value of force R is about 3.4 kN. It increases to 5.5 kN during the wheels' motion upside the inclined race and then drops down to 4.1 kN at the entry of the wheels on the vertical race. While moving up the race, the value of force R monotonically decreases to 2.4 kN at the end of the race.

3.3.2 Diagrams and calculations of forces acting on the top section of the retractable guidance system during its movement from resting to working position

Diagrams of forces acting on the top section of the retractable guidance system are presented in Figure 3.33. In the diagram marked as (a), forces acting on the guides in

Table 3.48 Values of the forces acting on the bottom section of the retractable guidance system during its movement from resting to working position

Marking of the Force	*Force Value*			
	Zero Moment	*Moment '1'*	*Moment '2'*	*Moment '3'*
Q	1.00 kN	1.01 kN	0.41 kN	0.68 kN
R	3.38 kN	5.45 kN	4.11 kN	2.42 kN
S	18.4 kN	17.1 kN	20.3 kN	20.1 kN

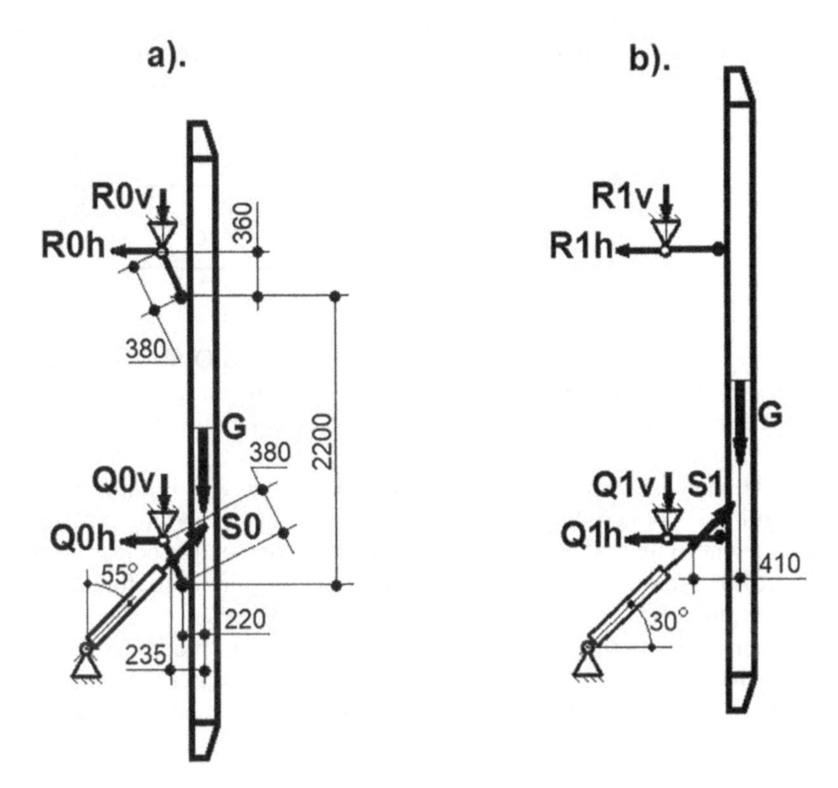

Figure 3.33 Diagrams of the forces acting on the top section of the retractable guidance system.

the initial moment, the so-called zero moment, and another one, labelled as (b) presents forces in the final moment – moment '1'.

In Figure 3.33, the following indications of forces were used:

- **G** – weight of the top section of the retractable guidance system,
- **Q0h**, **Q1h** – horizontal force acting on the bottom joint of the top section of the retractable guidance system, respectively, in moments zero and '1',
- **Q0v**, **Q1v** – vertical force acting on the bottom joint of the top section of the retractable guidance system, respectively, in moments zero and '1',
- **R0h**, R**1h** – horizontal force acting on the top joint of the top section of the retractable guidance system, respectively, in moments zero and '1',
- **R0v**, **R1v** – vertical force acting on the top joint of the top section of the retractable guidance system, respectively, in moments zero and '1',
- **S0**, **S1** – force caused by the drive of the top section of the retractable guidance system, accordingly in moments zero and '1'.

Diagrams presented in Figure 3.33 were drawn up according to Projekt techniczny (2018). Dimensions presented in Figure 3.33 were referred to the mentioned project, and the value of the weight G was calculated as equal to 18 kN.

Table 3.49 Values of the forces acting on the top section of the retractable guidance system during its movement from resting to working position

Moment	*Force Value*				
	Qh (kN)	*Qv (kN)*	*Rh (kN)*	*Rv (kN)*	*S (kN)*
0	37.0	18.0	14.4	0.0	62.8
1	60.5	18.0	1.80	0.0	72.0

Table 3.49 presents results of calculations of the forces Q, R and S values. The assumption was made that both inertial forces and friction acting on the top section of the retractable guidance system during its movement from resting to working position are negligibly small.

Table 3.49. shows that the values of the forces Q, R and S in the top section of the retractable guidance system during its motion from resting to working position are significantly different from each other, similarly to forces acting on the bottom section. The smallest force is Rv, which is equal to 0. In turn, the greatest value of the force is about 72.0 kN, which is the force S in the final moment of the guide's movement. Comparing to the initial value of this force, the final value is about 9.0 kN higher. The biggest rise characterizes force Qh, which is about 23.5 kN. The value of the force Rh decreases during the motion of the top section of the retractable guidance system from 14.5 kN to about 2.0 kN.

3.3.3 Diagrams and calculations of forces acting on the bottom section of the retractable guidance system during its operation

Diagrams of forces acting on the bottom section of the retractable guidance system during its operation are presented in Figure 3.34. In the diagram marked as (a), main frame of the bottom section in plane of shaft members receiving forces form the retractable guidance system is shown, and in the other diagram, labelled as (b), the bending of the frame in horizontal plane is presented. The assumption was made for the diagrams that in the construction of the retractable guidance system, temporary connections between its bottom section and the shaft chairing are disassembled. These connections were used for the purpose of installation of the retractable guidance system.

In Figure 3.34, the following indications of forces were used:

- **Pc**, **Pb** – horizontal force, respectively frontal (Pc) and side (Pb), acting on the main frame of the bottom section of the retractable guidance system during loading or unloading of the bottom level of the man-material cage on the level 960 m in the Leon IV shaft,
- **Pv** – vertical force acting on the main frame caused by movable swinging platform on the level 960 m,
- **Qb** – axial resistance force acting on the connection of the bottom shaft member with the shaft lining

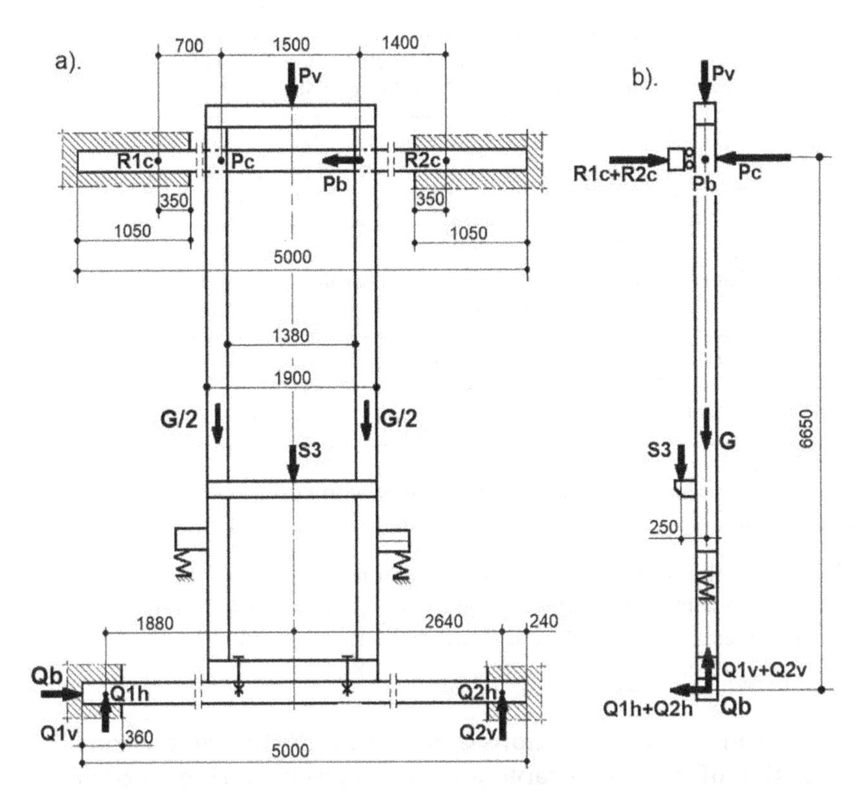

Figure 3.34 Diagram of forces caused by the bottom section of the retractable guidance system in its working position.

- **Q1h**, **Qh2** – vertical forces acting on the connection of the bottom shaft member with the shaft lining,
- **R1c**, **R2c** – horizontal forces acting on the connection of the top shaft member with the shaft lining,
- **S3** – force caused by the drive of the bottom section of the retractable guidance system, keeping it in the working position.

Diagrams presented in Figure 3.34 were drawn up according to Projekt techniczny (2018). Based on this project, the value of weight G and dimension shown in the figure were set. Values of forces Pc, Pb, Pv and S3, which were also presented in Table 3.50, together with the value of weight G, were adopted from the work of Płachno 2018b) and results of calculations made for forces acting on the bottom section of the retractable guidance system during its movement from resting to working position.

Table 3.51 consists of data obtained in calculations. The following assumptions were made for these calculations:

1. forces Pc, Pb, and Pv are static, according to Płachno (2018b),
2. peak values of each of the forces Q1v and Q2v are consistent with the orientation of the force Pb,

Table 3.50 Values of the forces acting on the construction of the bottom section of the retractable guidance system

G	Pc	Pb	Pv	S3
23 kN	20 kN	14 kN	100 kN	21 kN

Table 3.51 Values of the forces acting on the shaft lining caused by the bottom section of the retractable guidance system during its operation

Q1b	Q1h	Q2h	Q1v	Q2v	R1c	R2c
14 kN	0.46 kN	0.34 kN	111.1 kN	80.2 kN	16.1 kN	12.2 kN

3. peak values of each of the forces R1c and R2c are consistent with the orientation of the retractable guidance's element acting on the main frame.

Table 3.51 shows that the greatest forces acting on the lining of the Leon IV shaft and caused by the bottom section of the retractable guidance system during its operation are vertical forces acting on the connections of the bottom shaft member and shaft lining. Force Q1v is about 111 kN and Q2v about 80 kN.

3.3.4 Diagrams and calculations of forces acting on the bottom section of the retractable guidance system during its operation

Diagrams of forces acting on the top section of the retractable guidance system during its operation are presented in Figure 3.35. In the diagram marked as (a), the main frame of the top section in plane of shaft members transferring forces form the retractable guidance system to the shaft lining is shown, and in the other diagram, labelled as (b), bending of the frame in horizontal plane is presented. The assumption was made for the diagrams that are disassembled in the construction of the retractable guidance system temporary connections between its top section and the shaft chairing. These connections were used for the purpose of installation of the retractable guidance system.

In Figure 3.34, the following indications of forces were used:

- **Pc, Pb** – horizontal force, respectively, frontal (Pc) and side (Pb), acting on the main frame of the bottom section of the retractable guidance system during loading or unloading of the bottom level of the man-material cage on the level 960 m in the Leon IV shaft,
- **Q1h, Q1v, R1h** – vertical and horizontal forces acting on the main frame of the top section of the retractable guidance system during its operation,
- **R1c, R2c** – horizontal forces acting on the connection of the top shaft member with the shaft lining,
- **T1h, T2h** – horizontal forces acting on the connection of the shaft member supporting the main frame with the shaft lining,
- **T1v, T2v** – vertical forces acting on the connection of the shaft member supporting the main frame with the shaft lining.

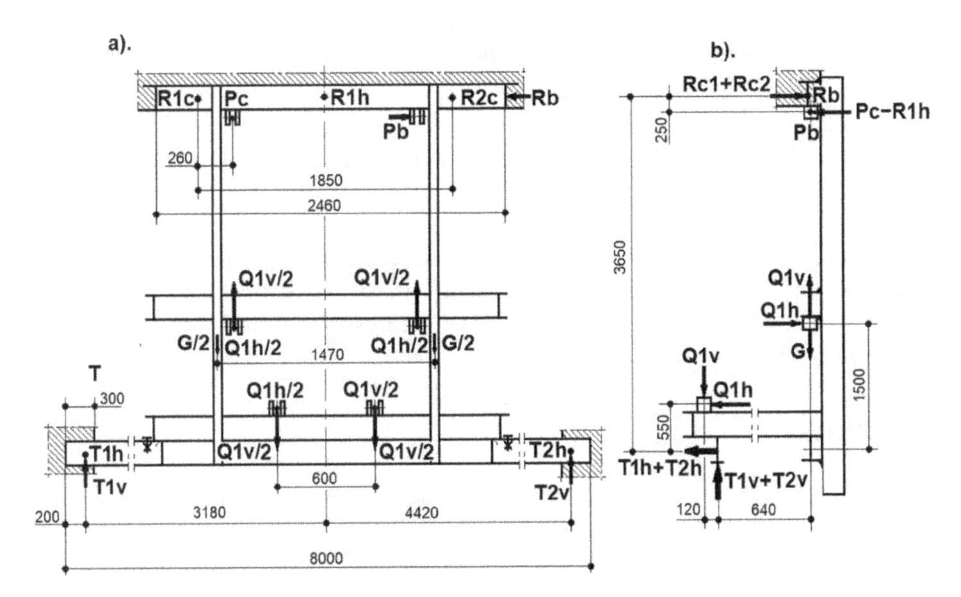

Figure 3.35 Diagram of forces caused by the top section of the retractable guidance system in its working position.

Table 3.52 Values of the forces acting on the construction of the top section of the retractable guidance system

G	*Pc*	*Pb*	*Q1h*	*Q1v*	*R1h*
19 kN	20 kN	14 kN	60.5 kN	18 kN	1.8 kN

Diagrams presented in Figure 3.35 were drawn up according to Projekt techniczny (2018). Based on this project, the value of weight G and dimension shown in the figure were set. Values of forces Pc, Pb, Q1h, Q1v and R1h, which were also presented in Table 3.52, together with the value of weight G, were adopted from the work (Płachno 2018b) and results of calculations made for forces acting on the bottom section of the retractable guidance system during its movement from resting to working position.

Table 3.53 consists of data obtained in calculations. The following assumptions were made for these calculations:

1. forces Pc, Pb, Q1h, Q1v and R1h are static, according to Płachno (2018b),
2. peak values of each of the forces R1c and R2c are consistent with the orientation of the retractable guidance's element receiving the force Pc.

Table 3.53 shows that the greatest forces acting on the lining of the Leon IV shaft and caused by the top section of the retractable guidance system during its operation are vertical forces acting on the connections of the top shaft member and shaft lining. Force Rb is about 14 kN, and forces R1c and R2c are both equal to 13.5 kN. In turn,

Table 3.53 Values of the forces acting on the shaft lining caused by the top section of the retractable guidance system during its operation

Rb	*R1c*	*R2c*	*T1h*	*T2h*	*T1v*	*T2v*
14 kN	13.5 kN	13.5 kN	6.5 kN	4.7 kN	11.1 kN	7.9 kN

the biggest values among the forces acting on the connection of the shaft member supporting the guidance's main frame and the shaft lining are vertical forces. Force T1v is about 11.5 kN, and T2v is equal to 7.9 kN. The greatest horizontal force acting on this connection is T1h, which is equal to 6.5 kN.

REFERENCES

ISO (1995) *Guide to the Expression of Uncertainty in Measurement.* ISO 1995. Switzerland.

Kamiński P., Prostański D. and Dyczko A. (2021) Test of the retractable guidance system installed on the level 960m in the Leon IV shaft in Rydułtowy Coal Mine, Poland, *IOP Conference Series: Materials Science and Engineering.* 1134 012001.

Płachno M. (2005) Nowe metody projektowania i eksploatacyjnej kontroli zbrojenia pionowych szybów górniczych, Monografia *Problemy Inżynierii Mechanicznej i Robotyki*", Wydawnictwo AGH, Kraków 2005 [In Polish].

Płachno M. (2018a) Doświadczenia poznawcze wynikające z diagnostycznych badań naprężeń zmiennych doznawanych przez cięgła nośne skipów górniczych o dużej ładowności. *Napędy i sterowanie* nr 1 2018.

Płachno M. (2018b) *Pomiar sił rzeczywistych oddziaływania klatki wielkogabarytowej w szybie Leon IV wraz z analizą mechaniczną układu i rozwiązań konstrukcyjnych ruchomego zbrojenia, analizą wytrzymałościową układu*, 2018 [In Polish, unpublished].

Polska Norma PN-G-46227. Szyby górnicze. Oszybia. Wymagania [In Polish].

Polska Norma PN-M-06515. Dźwignice. Ogólne zasady projektowania stalowych ustrojów nośnych [In Polish].

Projekt techniczny prowadzenia klatki wielkogabarytowej na poz. 960m (1000m). Projekt nr Ry-157/218 opracowany przez PBSz S.A. Górnicze Biuro Projektów w 2018 r., Autorzy: B. Chomański, K. Pyrek, P. Kamiński [In Polish, unpublished].

Rozporządzenie Ministra Energii z dnia 23 listopada 2016 roku w sprawie szczegółowych wymagań dotyczących prowadzenia ruchu podziemnych zakładów górniczych, Dz. U. z 2017 r. nr 1118 [In Polish].

Zaprojektowania uzbrojenia szybu Leon IV KWK Rydułtowy-Anna. Poziom 960m. Urządzenia przyszybowe ze sterowaniem. BSiPG, Katowice 2005 [In Polish, unpublished].

Chapter 4

Summary

Rope guidance of conveyances, commonly used in numerous underground mines all over the world, has numerous advantages connected with its assembly and operation, including economic benefits. It also has some disadvantages, and the greatest of them is a necessity for application of the stiff guidance in the vicinity of mine levels, which is necessary to provide safe operation of the hoisting system. It is also required by the Polish law. It is a serious issue in conditions of the Polish coal mining industry, where a hoisting system has to operate on many mine levels, because of the multi-level mine model.

The greatest inconvenience in the operation of the stiff guidance on mine levels is an entry of the conveyance into the guidance construction, because of lateral movements of the conveyance, amplified by the deceleration of the conveyance and turbulent airflow in the vicinity of mine levels. The requirement of the velocity reduction also negatively affects the effectiveness of the hoisting system.

To improve the effectiveness of the hoisting system operation with rope guidance of conveyances, the retractable guidance system was designed and then applied on the level 960 m in the Leon IV shaft in Rydułtowy colliery (currently ROW colliery, Rydułtowy department). The patented retractable guidance system allows the conveyance to move through the shaft station at the level 960 m with its full speed, if people or materials are transported to other levels, and if there is a need of transport to the level 960 m, it provides cage stabilization at the shaft station.

The described problem was neither analysed nor solved because rope guidance of conveyances is rarely used in Polish coal mines, mostly because of the listed issues. The innovativeness of retractable guidance system is the application of hydraulic cylinders to stabilize the conveyance in a safe manner. Additionally, any intervention in the construction of the conveyance is not necessary, as it was in historic solutions. The design of the retractable guidance system also allows to use swinging bridges on mine levels, which is important for the purpose of transport of materials to the level.

Wide commercial use of presented solution of the retractable guidance system might result in popularization of rope guidance also in mines, which were sceptical about it, both for historic and practical issues, among others – Polish coal mines. The use of rope guidance in mine shafts can result in numerous benefits, especially financial, for the mine.

As it was stated before, the presented solution of the retractable guidance system is a pioneering approach to the case of rope guidance in conditions of the Polish mining industry. It allowed to raise effectiveness of the hoisting system in Leon IV shaft of Rydułtowy mine, where it is successfully used. It can also help to raise levels of safety and effectiveness of other shafts of Polish underground mines.

DOI: 10.1201/b22695-4